Spectrum Management

With this definitive guide to radio spectrum management, you will learn from leading practitioners how spectrum can be managed effectively and made available both now and in the future.

All aspects of spectrum management are covered in depth, from the fundamentals of radio spectrum and technical and economic basics, to detail on methods such as auctions, trading, and pricing, and emerging approaches including shared and dynamic spectrum access and new ways of licensing. With the help of real-world case studies, you will learn how this knowledge comes together in practice, as the authors illustrate the role of spectrum in the wider economy and offer valuable insights into key future trends.

Authoritative and up-to-date, and bringing together the key technical, economic, and policy issues into one definitive resource, this is the essential guide for anyone working or studying in areas related to radio spectrum management.

Martin Cave is a regulatory economist who has worked extensively on telecommunications and spectrum issues. He is a visiting professor at Imperial College Business School and an Inquiry Chair at the UK Competition and Markets Authority. Previously he was a professor at Warwick Business School, BP Centennial Professor at the London School of Economics, and a member of the Spectrum Advisory Board of the UK regulator Ofcom.

Professor William Webb was instrumental in designing Weightless, a new global standard for wireless M2M communications, is CEO of the Weightless SIG, and has 17 patents pending or granted for the technology. He was a co-founder and CTO of Neul, a Cambridge start-up established to commercialize Weightless technology, was president of the IET during 2015, was a member of Ofcom's Spectrum Advisory Board (OSAB), and is on the board of TPRC. In 2005 he became one of the youngest ever Fellows of the Royal Academy of Engineering. He is also a Fellow of the IEEE.

Spectrum Management

Using the Airwaves for Maximum Social and Economic Benefit

MARTIN CAVE
Imperial College Business School and the Competition and Markets Authority

WILLIAM WEBB
Weightless SIG

CAMBRIDGE
UNIVERSITY PRESS

University Printing House, Cambridge CB2 8BS, United Kingdom

Cambridge University Press is part of the University of Cambridge.

It furthers the University's mission by disseminating knowledge in the pursuit of education, learning and research at the highest international levels of excellence.

www.cambridge.org
Information on this title: www.cambridge.org/9781107094222

© Cambridge University Press 2015

This publication is in copyright. Subject to statutory exception and to the provisions of relevant collective licensing agreements, no reproduction of any part may take place without the written permission of Cambridge University Press.

First published 2015

Printing in the United Kingdom by TJ International Ltd. Padstow Cornwall

A catalogue record for this publication is available from the British Library

Library of Congress Cataloguing in Publication data
Cave, Martin.
Spectrum management : using the airwaves for maximum social and economic benefit / Martin Cave, Imperial College Business School and the Competition and Markets Authority, William Webb, Weightless SIG.
  pages cm
Includes bibliographical references and index.
ISBN 978-1-107-09422-2
1. Telecommunication policy. 2. Radio frequency allocation – Economic aspects. 3. Radio frequency allocation – Social aspects. 4. Radio resource management (Wireless communications)
I. Webb, William, 1967– II. Title.
HE7645.C385 2015
384.54'524 – dc23
                                                                        2015014697

ISBN 978-1-107-09422-2 Hardback

Cambridge University Press has no responsibility for the persistence or accuracy of URLs for external or third-party internet websites referred to in this publication, and does not guarantee that any content on such websites is, or will remain, accurate or appropriate.

Contents

Preface		*page* ix
Acknowledgments		x
Plan of the book		xi
List of abbreviations		xii

Part I: Fundamentals   1

1   Spectrum management around the world   3
1.1   The uses of radio spectrum   3
1.2   Why spectrum needs managing   9
1.3   National spectrum regulation   11
1.4   International spectrum regulation   14
1.5   Differences across countries and regions   17
1.6   Global, regional, or national spectrum management?   18
1.7   Successes and challenges   21

2   The technical challenge   24
2.1   Introduction   24
2.2   Transmitting a radio signal   24
2.3   How signals propagate   28
2.4   Mechanisms of interference   33
2.5   Tolerance of interference   38
2.6   The need for regulation   40

3   The economic challenge: a basic primer on spectrum economics   42
3.1   Characteristics of spectrum as an economic resource   42
3.2   What is an efficient allocation of spectrum across uses?   44
3.3   A more realistic formulation of the problem   45
3.4   The broad range of modes of access to spectrum   46
3.5   Alternative ways of allocating and assigning spectrum   47
3.6   Conclusion   58

Part II: Economic management of spectrum   61

4   Using auctions to assign spectrum   63
4.1   Introduction   63

	4.2 Some types and effects of auctions	63
	4.3 Designing mechanisms to award spectrum licenses	65
	4.4 The spectrum auction process	68
	4.5 Auction theory	70
	4.6 Auction objectives	72
	4.7 Auction formats	74
	4.8 Combinatorial clock auctions	85
	4.9 Incentive auctions	88
	4.10 Conclusion	91
5	**Other aspects of spectrum auction design**	**94**
	5.1 Introduction	94
	5.2 Auction logistics	94
	5.3 Lot design	97
	5.4 Ensuring a competitive auction	98
	5.5 Auctions and downstream competition	103
	5.6 Can demand for unlicensed spectrum be accommodated in a spectrum auction?	109
	5.7 Conclusion	111
6	**Spectrum trading**	**113**
	6.1 Introduction	113
	6.2 Spectrum secondary markets	114
	6.3 Forms of spectrum trading	115
	6.4 Competition concerns and other objections to spectrum trading	117
	6.5 Spectrum trading in practice	120
	6.6 Concluding remarks	125
7	**Spectrum pricing and valuation**	**128**
	7.1 Introduction	128
	7.2 The separate components of spectrum prices	129
	7.3 Finding opportunity-cost prices: an initial approach	132
	7.4 Interrelations among opportunity-cost estimates	135
	7.5 Opportunity-cost spectrum pricing in practice	137
	7.6 Other pricing applications in practice	140
	7.7 Administrative prices and trading	143
	7.8 Conclusion	144

Part III: Sharing and other emerging approaches to spectrum management   147

8	**Spectrum sharing and the commons**	**149**
	8.1 Basic approach to commons	149
	8.2 The tragedy of the commons	152
	8.3 Restriction on usage in various bands	154
	8.4 The Ofcom Licence-Exemption Framework Review	157

		8.5 Summary	160
9		**Dynamic spectrum access**	162
		9.1 Introduction	162
		9.2 Approaches to dynamic access	163
		9.3 Licensed shared access	167
		9.4 Unlicensed shared access	168
		9.5 Advantages and disadvantages of shared access	172
		9.6 Example 1: TV white space	174
		9.7 Example 2: US 3.5 GHz band	184
		9.8 Example 3: government sharing	185
		9.9 In conclusion: the need to increase flexibility	187
10		**Controlling interference: licensing and receivers**	192
		10.1 Introduction	192
		10.2 Spectrum usage rights	192
		10.3 Receiver standards	197

Part IV: Case studies and conclusions — 205

11		**The struggle for the UHF band**	207
		11.1 The issues at stake	207
		11.2 Broadcasting, the digital switch-over, and current trends	208
		11.3 Broadcasting technical options	212
		11.4 Mobile data, national broadband plans, and spectrum management	215
		11.5 Smartphones and the data crunch	217
		11.6 Resolving noneconomic valuation issues	219
		11.7 Finding an efficient allocation for the 700 MHz band	221
		11.8 The struggle for the UHF band: the options	226
		11.9 Possible outcomes	228
		11.10 Implications for spectrum management	229
12		**Public-sector spectrum use**	231
		12.1 Introduction	231
		12.2 Differences between commercial and public-sector use	232
		12.3 A program of reform of public spectrum use	234
		12.4 An example of public-sector spectrum reform: the UK	239
		12.5 Conclusion	240
13		**Spectrum and the wider economy**	241
		13.1 Introduction	241
		13.2 Spectrum, spectrum-using services, and their impact on welfare	241
		13.3 Effects of spectrum-using services on GDP and employment	243
		13.4 Effects of spectrum-using services on productivity	244
		13.5 Conclusion	248

		13.6	Annex	249
14		**Where next?**		**252**
		14.1	Trends	252
		14.2	Our agenda to improve spectrum use	254
		14.3	In conclusion	256

About the authors   258
Index   260

Preface

We published our previous book, *Essentials of Modern Spectrum Management*, in 2007. Much in the world has changed since then, with the explosion in demand for mobile data, the emergence of dynamic spectrum access and other sharing approaches, and the deployment of new auction techniques. Some of the developments that were promising in 2007, such as ultra-wideband, have not yet delivered, while others, such as television white spaces, are now being pioneered.

When we decided it was time for a new edition, we concluded that the changes required to the 2007 version were so extensive as to merit a completely new approach. Hence this book, which aims to cover the major issues relating to the technologies, economics and practices of using and managing spectrum, to consider different approaches, to look ahead, and to make recommendations for future spectrum management.

Acknowledgments

The authors are grateful to Drs Chris Doyle and Rob Nicholls, who contributed to Chapters 4 and 5, and to Dr. Leo Fulvio Minervini, who contributed Chapter 7. Drs. Minervini and Nicholls and Adrian Foster read the manuscript and gave the authors helpful comments. Thomas Welter kindly furnished us with certain information on spectrum policy in France. Neil Pratt contributed to Chapter 11, and Graham Louth also helped us with advice and comments.

Plan of the book

This book consists of four parts:
   The first is a primer, designed to ensure that all readers have sufficient knowledge to tackle the material in the rest of the book. It covers spectrum management fundamentals, technical issues, and basic economics.
   The second covers conventional economic methods of spectrum management, such as auctions, trading, and pricing, which have been evolving for a decade or more.
   The third looks at spectrum management approaches which we believe will become more prominent in future, including shared and dynamic spectrum access and new ways of licensing based on interference caused.
   The fourth looks at some case studies and issues. It uses the UHF TV band to illustrate a number of principles from earlier chapters, considers approaches that might be adopted in the public sector in international spectrum management, and examines the role of spectrum in the wider economy. Finally, it contains our projection of trends and the key agenda which we think needs to be tackled.

Abbreviations

3GPP	Third Generation Partnership Project
ACL	adjacent channel leakage
ACMA	Australian Communications and Media Authority
ACS	adjacent channel selectivity
AGC	automatic gain control
AGL	above ground level
AIP	administered incentive pricing
ANFR	Agence nationale des fréquences (France)
ATC	ancillary terrestrial component
AWS	advanced wireless services
BAS	broadcast auxiliary service
BFWA	broadband fixed wireless access
CCTV	closed circuit TV
CEO	chief executive officer
CEPT	Central European Post and Telecommunications
CMA	cellular market area
CW	continuous wave
DAB	digital audio broadcasting
DECT	digital European cordless telephone
DoD	Department of Defense (US)
DSA	dynamic spectrum access
DTT	digital terrestrial television
DVB	digital video broadcasting
EA	economic area
EBU	European Broadcasting Union
ECC	European Communications Committee
EIRP	equivalent isotropic radiated power
eMBMS	evolved multimedia broadcast multicast service
EMC	electromagnetic compatibility
ETSI	European Telecommunications Standards Institute
EU	European Union
FCC	Federal Communications Commission
FDD	frequency division duplex
GAA	general authorized access

List of abbreviations

GDP	gross domestic product
GHz	gigahertz
GPS	global positioning system
GSM	global system for mobile communications
HTHP	high-tower high-power (transmitter site)
ICT	information and communications technology
IEEE	Institution of Electrical and Electronic Engineering
IET	Institution of Engineering and Technology
IoT	Internet of Things
IPTV	Internet protocol TV
ISD	inter-site distance
ISM	industrial, scientific, and medical
ITU	International Telecommunication Union
kHz	kilohertz
LEFR	Licence-Exemption Framework Review
LSA	licensed shared access
LTE	long-term evolution (of cellular technology)
LTLP	low-tower low-power (transmitter sites)
M2M	machine-to-machine
MCL	minimum coupling loss
MED	Ministry of Economic Development (New Zealand)
MFN	multifrequency network
MHz	megahertz
MIMO	multiple-input multiple-output (antennas)
MNO	mobile network operator
MPEG	Motion Picture Experts Group
NAB	National Association of Broadcasters
NATO	North Atlantic Treaty Organization
NRA	national regulatory authority
NTIA	National Telecommunications and Information Administration
OSAB	Ofcom Spectrum Advisory Board
PCAST	President's Council of Advisors on Science and Technology
PCS	personal communications services
PFD	power flux density
PFWA	public fixed wireless access
PMR	private mobile radio
PMSE	program making and special equipment
PPDR	public protection and disaster relief
PSB	public-service broadcasting
PVR	personal video recorder
RET	revenue equivalence theorem
RFID	radio frequency identification
RSC	Radio Spectrum Committee

List of abbreviations

RSPG	Radio Spectrum Policy Group (of the EC)
RSPP	Radio Spectrum Policy Programme
SAA	simultaneous ascending auction
SAS	spectrum access system
SDARS	satellite digital audio radio service
SFN	single-frequency network
SIG	special interest group
SIM	subscriber identity module
SLC	significant lessening of competition
SMR	specialized mobile radio
SMRA	simultaneous multiple-round auction
SNR	signal-to-noise ratio
SUR	spectrum usage right
TDD	time division duplex
TNR	Transfer Notification Register
TVWS	TV white space
UHDTV	ultra-high-definition TV
UHF	ultra high frequency
UN	United Nations
UWB	ultra-wideband
VHF	very high frequency
W	watt
WCS	wireless communications service
WRC	World Radio Conference

Part I

Fundamentals

1 Spectrum management around the world

1.1 The uses of radio spectrum

1.1.1 Introduction

The use of radio spectrum is at the heart of almost all aspects of daily personal, business and government activities. Life without the services reliant upon spectrum would be unthinkable – either no or very limited TV, radio, Internet, air travel, mobile phones, and much, much more. In this section, we provide an overview of the key uses of radio spectrum, their current allocations, and their likely future needs. As an illustration, Figure 1-1 shows the key uses of the spectrum in the UK and their current split of allocations across broad frequency bands.

In interpreting the figure (and recalling that it refers only to the UK) it is, first, worth noting that the usage of all frequency bands will add to greater than 100%. This is because much of the band is shared, and in the chart, if two applications both shared, say, 10% of the band, both would be considered to be using this 10% since it is typically very hard to divide up shared utilization. On this basis, some uses appear overstated: for example, program making and special equipment (PMSE), which is broadly the use of wireless microphones and cameras, appears to have more access to spectrum below 1 GHz than mobile telecommunications, but all the PMSE allocation is shared with broadcasting and PMSE needs to work around broadcasting, giving it little "real" allocation in practice.

In the figure the use is split into four different frequency bands. The advantages and disadvantages of different frequencies are discussed in more detail in the next chapter. Suffice it to say here that the bands below 6 GHz are considered much more valuable than those above and that the band below 1 GHz is especially useful where long-range propagation is required. We now turn to each of the sectors to provide a brief overview.

1.1.2 The public sector

The public sector encompasses all governmental (local, regional, and national) use, which is very varied. By far the largest users in this sector are defense and aeronautics. Other significant users are the emergency services and entities such as maritime safety (e.g. the coastguard) and scientific research. Thus, unlike the categories identified below, public-sector spectrum is put to a multitude of uses.

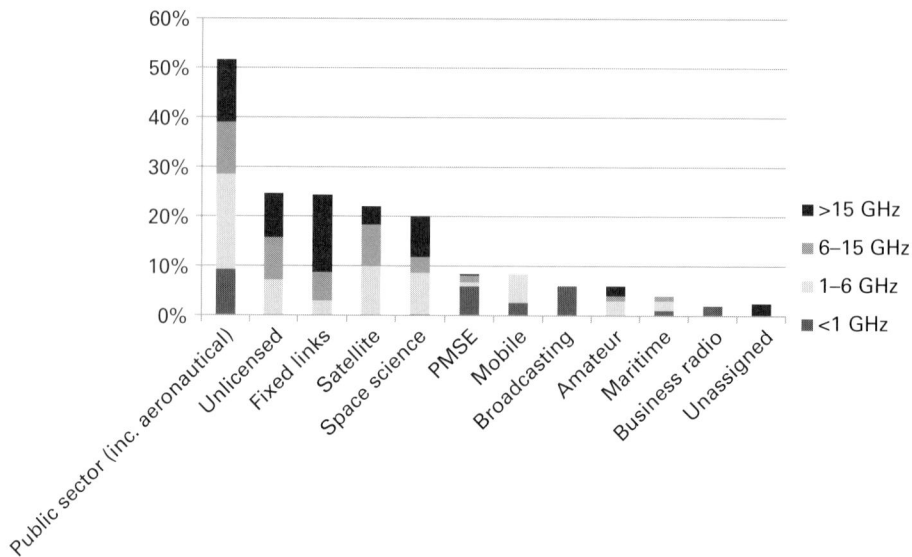

Figure 1-1. Uses of spectrum in the UK. Source: [1].

In most countries, defense services are a major user of radio spectrum. In some ways, this is a reflection of all the other uses – defense has its own "mobile" radio system, its own aeronautical system, its short-range systems, and so on. The actual use of spectrum is often confidential in terms of both the applications each band is used for and the level of usage. Over the last few decades defense use of spectrum has been slowly declining in the advanced economies, as defense needs domestically decline and the value of commercial use has risen. However, in recent years new technologies, such as pilotless drones which require video links, suggest that usage may need to grow to meet these needs. Understanding just how much spectrum should be set aside for defense and policing is extremely difficult and covered in detail in Chapter 12.

Aeronautical use of spectrum is predominantly for radar navigation. There are many different radar systems, from long-range radars that monitor the entire airspace across a country to approach radars at airports and even radars that monitor runways for debris. Planes also use radar for detection of terrain, other planes, turbulence and adverse weather. The accuracy of a radar system is related to the bandwidth of spectrum used and, with requirements for high accuracy, for most aeronautical applications a large amount of bandwidth is dedicated to these radars. Their use generally needs to be standardized internationally so that planes can use the same systems around the world. Other aeronautical applications include the radio links to pilots which sit in the VHF band just above sound broadcasting, data links for passenger in-flight applications which tend to use satellite systems, and radio altimeters which provide height-above-ground information to pilots. Broadly, aeronautical use of spectrum is static, making any changes to the allocation extraordinarily difficult.

The emergency services use spectrum for their own private radio systems. This is broadly the equivalent of a cellular solution for police, fire, and ambulance. By owning it themselves the emergency services have greater control over coverage, availability, reliability, and features. At present, the emergency services are looking for additional frequencies so that they can increase the data capacity of their system, although it is far from clear whether they will be granted anything globally (there are some exceptions such as in the US[1]). This category of use is sometimes termed public protection and disaster relief (PPDR). It has also been suggested that such users could share with private users of mobile communications.

1.1.3 Unlicensed use

Unlicensed use (sometimes known as "spectrum commons" or as "license-exempt") is a broad set of applications, the best known of which are Wi-Fi[2] and Bluetooth. Unlicensed bands are somewhat like public parks – anyone is allowed to use them as long as they follow the rules of entry; this is discussed in detail in Chapter 8. While unlicensed use appears to have a large allocation of spectrum, it has virtually none below 1 GHz, and much of its allocation is either shared or in bands considered "junk" because of interference from other systems. Unlicensed use appears to be growing rapidly with ever-greater deployment of Wi-Fi, and many new applications such as machine-to-machine (M2M)[3] communications are considering unlicensed bands.[4] Regulators are now looking at whether unlicensed access could be granted into licensed bands on an opportunistic basis – known as dynamic spectrum access or "white-space" access. Deciding how much spectrum should be unlicensed is very difficult as generally market mechanisms such as auctions cannot be used and the future use of a band is hard to predict. Unlicensed bands are often seen as fertile areas for innovation as new technologies and applications can be deployed at low cost and without the need to gain a license [2].

1.1.4 Fixed links

Fixed links transmit data from one point to another using thin beams of radio energy. They are used in places where good-quality copper or fibre lines are unavailable or uneconomic. One major use is "backhaul" from mobile phone cell sites, linking the radio equipment at the cell tower back into the core network of the operator. Other applications include backup links for major cables and provision of telecommunications to remote

---

[1] The US has tried for many years to engineer a public–private partnership for the delivery of PPDR using the auction mechanism as a tool. The 700 MHz auctions had a reserved component ("D block") with obligations on the private owner to provide PPDR services, but this was unsold in the initial auction. Shared solutions now appear to be emerging through the "FirstNet" initiative.

[2] In 1985 the FCC in the US released the 2.4 GHz ISM band (the industrial, scientific, and medical radio bands reserved internationally for the use of industrial, scientific, and medical purposes other than telecommunications) for unlicensed access. This led to innovations giving rise to the 802.11 standard otherwise known as Wi-Fi, a low-powered, radio-based local-access network. Bluetooth is a related standard using the ISM band to enable personal-area networks.

[3] Sometimes known as the Internet of Things (IoT), M2M is the transmission of data to and from machines such as smart meters and parking sensors.

[4] For example, see the Sigfox technology deployed in 868 MHz or the Weightless technology deployed in "TV white space."

villages. Fixed-link antennas often look like satellite dishes but are pointed horizontally. Because directional antennas result in narrow, focussed beams of radio energy, the radio signal travels further for a given transmitted power. This allows fixed links to operate in higher-frequency bands where signals propagate less well but there is more spectrum and it is less expensive. The demand for fixed links in advanced economies is approximately static, although there is something of a trend away from longer-range links towards short-range solutions for small cells in cities. This is relatively easy to manage as the shorter links can be moved to ever-higher frequency bands where there is currently much available spectrum. Also, because of their thin beams of energy, fixed links rarely interfere with each other, and other applications can often share the same bands.

1.1.5 Satellite

Satellites are used for a wide range of applications. One is communications to remote users, providing voice and data services to ships and increasingly aircraft over sea. Satellite operators such as Inmarsat are periodically updating the data rates and capacity of their system by launching next-generation satellites. Another major use is satellite TV broadcasting, with households in many countries reliant on satellite for TV reception. Global positioning systems such as GPS and the new Galileo constellation are also very valuable users of satellite radio spectrum. Satellite solutions are also used for backhaul to remote sites, for news-gathering in isolated areas, for emergency communications during disaster situations and more. Satellite requirements for spectrum are under continual pressure, especially in the lower-frequency bands, and generally satellite use is progressively being moved to bands above 6 GHz. Satellite allocations are managed at a global level since the coverage from a single satellite can typically be greater than the size of most countries.

1.1.6 Space science

Another use of satellites is monitoring for scientific purposes. This is often known as "earth observation satellite" and includes satellites for meteorological purposes, for measuring parameters such as sea temperature or various emissions, for scientific missions into outer space, and so on. These systems often need to use specific frequency bands where the monitoring equipment operates best, and need these bands to be particularly "clean" of interference due to the sensitive nature of their instruments. There is slowly growing demand in this area, often for lower-frequency spectrum that is also in high demand from other applications.

1.1.7 Wireless microphones and cameras

In the world of security, entertainment and the media more generally, and sport, the demand for wireless microphones and wireless cameras has grown markedly in recent years. These are used in program making, but wireless microphones are also used very widely elsewhere. These applications require relatively little spectrum because the range involved is typically short, but they can require interference-free links – for example interference on the wireless microphones used in major stage shows cannot be tolerated. Wireless microphones typically share spectrum with others, often broadcasters. However, this secondary-user status makes them vulnerable whenever the primary use is changed and as a result they are sometimes moved to different frequency bands. The demand for wireless links for program making is growing with the increasing use of wireless cameras embedded in sports equipment and similar, but the spectrum availability is, if anything, declining. The user base is diverse, making it difficult to acquire spectrum through vehicles such as auctions, but the reliability requirements make unlicensed spectrum generally unsuitable. It is rather a "problem child" for which regulators have yet to find a satisfactory long-term solution. Major events like the FIFA World Cup and IOC Olympics give rise to huge spikes in demand for spectrum to support the uses discussed here. To accommodate such demand sometimes regulators have temporarily withdrawn spectrum from other users.

1.1.8 Mobile telecommunications

Mobile, or cellular, telecommunication is one of the most visible users of spectrum. It comprises multiple generations of technology (2G, 3G and 4G/LTE) across many different frequency bands in the 300 MHz–3 GHz frequency range. It is generally considered to be the most economically valuable use of the radio spectrum and demand in advanced economies continues to rise as numerous new data-intensive Internet-based applications emerge. Because of rising demand for data fueled by the Internet and smartphone usage, regulators have had to refarm bands that were previously allocated to uses such as broadcasting. Over the years, ever more spectrum has been found for mobile, most recently at 800 MHz and 2.6 GHz (in Europe) and 700 MHz–2.3 GHz (in the US), and there is likely to be ongoing pressure for further refarming. Spectrum licenses assigned for use by mobile service providers are generally awarded through auction processes that often raise enormous sums for government. To date the largest sum raised in a single spectrum auction was that associated with the German 3G auction held in the year 2000, which raised almost €51 billion. Between 1994 and 2015 the US government has sold spectrum in over 96 different auctions, many of them awarding licenses to mobile operators, including the AWS-3 auction in 2015 which netted the US government $45 billion [3].

1.1.9 Broadcasting

Broadcasting encompasses radio at a number of frequencies, the main one being VHF radio (around 90–110 MHz) and TV at UHF (around 170–240 MHz and 470–790 MHz). The bandwidth required for radio is much smaller than that for TV broadcasting. Terrestrial TV transmissions typically take place from high towers and at very high power levels, providing coverage across large areas of the country, and causing substantial interference. In the last decade TV transmissions have been upgraded from analogue to digital systems in a process known as digital switch-over, which has enabled more TV channels and high-definition TV broadcasts. In an ideal world, broadcasters would like massively more spectrum, to deliver more channels and to enable more high-definition and ultra-high-definition broadcasting, but it is clear that this is not possible in the current frequency bands. This has resulted in the use of satellite and cable systems to deliver much of the TV content, which has left question marks over the future of terrestrial TV broadcasting in UHF (often known as digital terrestrial television – DTT), especially as there is great demand for this spectrum from mobile networks and others. The debate over the balance between broadcasting and other uses is a critical one for the spectrum community and is discussed in detail in Chapter 11. Because of the large range of TV transmitters, much TV usage is planned on an international basis, for example across the whole of Europe, making any change in this area complex and bureaucratic.

1.1.10 Amateur use

Amateur radio users (sometimes known as "radio hams") are granted access to some spectrum for free to allow them to experiment and indulge a hobby. Historically, such users have been valuable in monitoring transmissions from ships in distress and similar, but with better communications this use is in decline. They have also pioneered new technologies, but with ever-increasing investment needed to realize gains it is unclear whether this is likely in the future. Despite this, their entitlement continues, typically being recognized at an international level. Occasionally there is a desire to reclaim some of the bands set aside for amateurs, but as many of these are at higher and less valuable frequencies this is relatively rare.

1.1.11 Maritime use

Maritime use is for links to ships within communications distance of the shore – typically some 30 miles or so. This covers port use. This is a stable requirement with an increasing demand for data transmission, although some of this can be met using commercial cellular and satellite systems. Maritime radio frequencies are typically harmonized globally.

1.1.12 Business radio

Business radio, also known as private mobile radio (PMR), is used by companies such as taxis, and within airports, universities, business parks, and the like. A business radio system typically comprises a single transmitter, often located at the head office of the company, and radios in vehicles or carried by individuals. It has the advantage of low cost and simplicity, and the "broadcast" nature of transmissions such as "Is any taxi free in the vicinity of location X?" can be useful for some purposes. It was long thought that business radio would be progressively replaced by cellular solutions, but demand has held approximately constant, or is even growing slowly in some countries. Systems are often voice-based, although there are also low-data-rate capabilities in some cases.

1.1.13 Unassigned spectrum

Finally, some spectrum is unassigned, either pending assignment through an auction or because a decision on its use cannot currently be made and is better deferred. Regulators aim to keep the amount of unassigned spectrum small since it represents an unused resource. On the other hand, reserving spectrum for future use keeps options open for the future, and this can serve a valuable purpose.

1.2 Why spectrum needs managing

We have seen that radio spectrum supports many different uses and demand across these uses is changing and variable. Furthermore, the technology supporting radio-based services is evolving, and sometimes it eases pressure on demand for spectrum access but often exacerbates demand. With such changes taking place there is a need to manage access to frequency bands. Additionally, leaving frequency bands unchecked and allowing free access would give rise to the potential for much harmful interference, potentially obstructing services and having enormous negative costs for economies. For these reasons radio spectrum has been closely managed for decades, both at a global level through the ITU (a body within the UN) and at a national level.

Spectrum (radio frequency bands below 300 GHz) can be viewed as a natural resource, the value of which (unlike that of land, air, or water, but like that of oil and gas) escaped the notice of the human race for most of its existence. Although the amount of frequency in a given area is fixed, there are in principle limitless ways of dividing its use. This suggests that supply of spectrum is not an issue (unlike oil or gas). However, from a practical perspective spectrum is limited or finite as modern radio engineering systems require a degree of exclusive access at any given moment in time. Across the radio spectrum propagation properties vary and this means that different bands are better suited to different uses.

For much of the period in which spectrum's existence has been recognized, its importance has been marginal, recognized mainly by experts in physics, aeronautics, defense, and security. After the 1914–18 First World War, it began to be used to carry mass-point-to-multi-point radio broadcasts and, after the Second World War, television broadcasts. What finally brought spectrum to prominence was the diffusion of mobile communications, in the form first of voice and now, increasingly, of data. It is estimated that in 2015 there are about 7 billion mobile connections in existence, by means of which about 3.5 billion inhabitants of the planet have mobile voice connectivity.

Soon after the development of the mobile industry, finance ministries began to understand the value of spectrum through a series of auctions for spectrum licenses occurring at the turn of the century. The revenues received in the year 2000 by the German and UK governments in particular, from their 3 mobile licenses, drew attention to the scarcity value of spectrum suitable, and available, for mobile communications. Makers of public policy are now also aware of the contribution which wireless voice and broadband services can make to GDP growth. In the private sector, spectrum trades between operators, especially in the US, have generated billions of dollars of revenue, in addition to government auction revenues.

As spectrum became increasingly scarce the methods chosen to allocate it have developed. About a century ago, there was a short interlude of open access, in which users generally coexisted without interference. But governments in many countries quickly took control of spectrum for reasons of defense and security (particularly at the time of the 1914-18 world war), and then over the years eked it out for other uses, by means of administrative or "command-and-control" methods. This typically entailed a requirement placed on a would-be private or commercial spectrum user to seek the grant of a highly specific spectrum license, which would detail the use to which the spectrum would be put, the power levels and often the location of transmission equipment, permission often being granted for a short but renewable period. If there was competition to provide a service, the government would decide which applicant was the most worthy. Public-sector users, such as defense forces, usually found it much easier to get access to spectrum: they had the government's ear and faced no competing organization delivering the same services (and so requiring access to the same spectrum).

Despite the built-in resistance to innovative spectrum uses which this system embodied, it was a fairly effective way of dealing with spectrum assignment in a period of spectrum abundance. However, as with other resources, growing demand can trigger a search for additional resources and alternative modes of allocation. This is exactly what has happened in the last decade or so. A symptom has been widespread discussion of what is known as the "coming spectrum crunch" – a prospective shortage of spectrum to meet the burgeoning needs of mobile broadband. The expectation of this event has elicited a number of significant responses in spectrum management:

1. Searching in higher bands for capacity to deal with growing demand, initially at around 3.5 GHz but more recently even in bands above 20 GHz, such as the 60 GHz unlicensed band.
2. Questioning whether the traditional administrative or "command-and-control" approach is adequate to the task of making the best use of spectrum in an era of shortage and accelerating technical progress. The first fully developed explicit proposal for use of an alternative mode of allocation – use of a market for spectrum – was made in 1959 by Ronald Coase, but it took another 30 years for market instruments, initially in the form of spectrum auctions, to begin to take hold.
3. Looking for ways for bands to be shared among multiple users and multiple uses. This has been a most prominent feature of the past decade or so. Technical developments have made it possible for multiple users to have access to the same band without incurring the interference which, formerly, would afflict both transmissions. Sharing has also become a means of dealing with the "overhang" of public-sector spectrum use, which arose from the privileged position of those users of spectrum in the days of spectrum abundance. If it is difficult to persuade public-sector users to give up their assignments, it might be easier to make them share them. And data on spectrum use do reveal extensive underuse of public-sector capacity, in some cases even in peak hours.

The basic economic principle of efficient resource allocation for an input such as spectrum is that a band with broadly the same functional capability (for example a band capable equally of enabling both broadcasting and mobile communications services) should be divided between any two uses in such a way that the change in benefit to society (measured in monetary terms) delivered by assigning an additional megahertz of spectrum to each service should be the same. Thus if an additional megahertz of spectrum applied to mobile communications delivered a benefit of $2 million a year when withdrawal of a megahertz of spectrum from broadcasting led to a decline in benefit of $1 million a year, it would be an indication that too much spectrum was being assigned to broadcasting.

This simple principle needs some discussion. First, with rapidly changing demand, and given the time it takes to shift spectrum from one use to another, it will be impracticable continuously to be shifting spectrum from one use to another to satisfy the above condition at each moment. The realizable goal is to achieve the desired relationship approximately over time. Second, the benefits associated with spectrum-using services consist not only of the willingness of the individual users to pay for the relevant service. Wider social goals and further economic benefits accruing to others than immediate consumers should also be taken into account. Third, the condition stated above assumes that we know what services are available. Another key goal of spectrum management is precisely to create conditions in which new services can be made available to users.

Data discussed in a later chapter suggest that spectrum has made, indirectly, a growing contribution to GDP. Asking how much worse off we would be if spectrum did not exist probably makes no better sense than asking how much worse off we would be without water, oil and gas, or any other natural resource: there is no indication that spectrum is likely to go away. But the findings discussed below, to the effect that the spectrum contributes around 3% of GDP do give an order of magnitude to the importance to getting spectrum management right. Thus if our current probably inefficient regime of spectrum management can deliver figures like those quoted above, what would be the payoff to a step change in its efficiency? But perhaps it is unrealistic to talk in terms of achieving an efficient allocation – rather we should emphasize that the current system is in some sense "third best" and we can improve upon this by allowing greater flexibility via market-based instruments to achieve a superior second-best outcome.

1.3 National spectrum regulation

Spectrum is largely managed at a national level. While there are international bodies and agreements, discussed in the next section, the legal rights to grant licenses and prosecute those who break laws relating to spectrum use are invariably held by national governments. Each government appoints a national regulatory authority (NRA) which is responsible for the management of the spectrum. In some cases the necessary powers are embedded within government as a part of a relevant government department (although which department can sometimes be unclear given the breadth of use of spectrum). In other cases, such as Ofcom in the UK, they are set up as stand-independent regulatory agencies, with substantial autonomy within the terms of reference [4] set for their operation.

The breadth of remit of the NRA also varies. In some cases it is purely spectrum-related; in others it is a "converged" regulator covering other activities such as broadcasting and telecommunications regulation. The argument for convergence is that otherwise, for example, a broadcaster might find itself getting conflicting regulation from one entity seeking to reduce its spectrum usage and another demanding greater content or coverage. By bringing both decisions under one roof, coherent decisions can be made that take all the relevant factors into account.

The functions of the NRA can broadly be grouped into

- determining and changing the use of spectrum,
- assigning cleared spectrum,
- enabling change of ownership and use of encumbered spectrum,
- managing unlicensed spectrum,
- policing spectrum use and resolving interference issues,
- conducting forward-looking studies to enable spectrum allocation to adapt, and
- partaking in international forums.

Each of these is introduced briefly below – and collectively they form the content of this entire book.

Determining and changing the use of spectrum. Spectrum bands are often allocated to particular uses – such as broadcasting. Over time the optimal use of the spectrum will change as new applications emerge and new technologies enable greater efficiency and different approaches. Firms in some jurisdictions are entitled under their license conditions to change the technology which they use to provide a given service – for example to switch from 2G mobile technology to 4G. It might also be possible for the firms in a spectrum marketplace to change the use itself, with one user group (e.g. broadcasters) selling its spectrum to a different group (e.g. mobile operators). We discuss the extent to which this might be possible in the future in later chapters, but for the moment most change of use requires substantial regulatory intervention. For example, in the recent change of use of the 800 MHz band from broadcasting to cellular there was first a European-wide agreement that this would happen,[5] followed by widespread co-ordinated replanning of the broadcast band, and national project management to move transmitter frequencies and help viewers retune their receiving equipment (TVs, set-top boxes, etc.) as needed. This also required moving of wireless microphones in some countries using a regulator-managed scheme to reimburse the costs of affected users [6].

Assigning cleared spectrum. Once spectrum is cleared (or even partially cleared) the regulator needs to decide what use to put it to. A first key decision is whether it should be unlicensed or licensed. If unlicensed then the rules of access need to be set. If licensed

---

[5] This was performed by the CEPT [5].

then the means of distribution need to be defined, generally using an auction. Auctions are subjects of substantial work and a large part of this book is devoted to the issues and complexities. Regulators typically retain teams of engineers (to set the technical license conditions), lawyers (to structure the legal framework), and economists (to set the auction format) during an auction – with the total size of a team for a large country sometimes exceeding 100 people once both direct employees and consultants are taken into account.

Enabling change of ownership and use of encumbered spectrum. There may be cases where spectrum ownership can be changed by the market. This is often called trading and is the sale of a license by one party to another. There are many options, including leasing, and full or partial sale. It is also important whether the use of the spectrum can be varied. These issues are discussed in more detail in subsequent chapters. Regulators typically first need to set appropriate regulations which enable this activity; this has not been done in all countries. They then often provide oversight to the process to ensure that trades are appropriate in relation to interference which might occur or competition problems which might arise.

Managing unlicensed spectrum. Unlicensed spectrum can need periodic management. This could involve changing the rules of access for reasons such as enabling new technologies or because congestion was occurring.

Policing spectrum and resolving interference. A key duty of the regulator is to ensure that users do not suffer "harmful interference." This is an ill-defined term [7] but broadly users will report to the regulator when their licensed use is being disrupted by interference and will expect the regulator to investigate and terminate unlawful use. Regulators typically maintain a "field force" of monitoring stations and vehicles equipped with direction-finding systems such that they can track down unwanted transmissions. These teams are often made available 24/7 since some interference, affecting, for example, air traffic control systems, can raise safety-of-life concerns. Where interference is found the teams may have powers to force entry into premises and seize equipment, or may work with the national police force to take appropriate action. A continual source of interference is "pirate" radio transmissions in the VHF radio band.[6] While the transmitters are relatively easy to track down, it is also easy for the parties involved to hide the studio and individuals responsible in a different location and then re-equip the transmitter site within hours. In a country like the UK a field force team of 30 to 50 trained engineers might be deployed.

Conducting forward-looking studies to enable spectrum allocation to adapt. Ideally, spectrum managers should be a decade ahead of new technologies and applications. This is because it can often take a decade to agree to clear a band and reassign it. Without clear regulatory intent to allow a new technology it will be difficult for entrepreneurs to raise venture capital. Hence the regulator can act as a brake on innovation and new technology. This was at least partly the case with ultra-wideband, a new technology that promised to provide very high data rates over short

---

[6] See Ofcom's guidance on tackling pirate transmissions at [8].

distances and to replace cables.[7] Being so far ahead of the technology curve is very difficult but if the regulator conducts studies, engages stakeholders through consultation, and frequently reappraises its strategy, it can be alert to new ideas, give the community some guidance as to its thinking, and share best practice across all the interested parties. A regulator might wish to spend between $1 million and $10 million a year on activities of this sort depending on its size. This spending is often seen as a good investment in that it will typically lead to auctions that net very much greater sums for the Treasury.

Partaking in international forums. As discussed in more detail below, there are many relevant international bodies, and regulators need to set aside resources for engaging with these. Often, these will be specialist individuals who understand the myriad complexities of international spectrum management.

Finally, it is worth commenting on the culture of regulation. It is natural, and broadly right, for any regulator to be risk-averse. Firms investing in long-lasting assets place a high value on regulatory certainty. Often the cost of destabilizing the status quo is much greater than the benefit from a new approach. So regulators tend to work slowly, seeking near certainty in their analysis rather than aiming for time or cost deadlines, unlike the commercial world where in some cases it is appropriate to make some quality sacrifices to deliver on time.

However, as suggested above, where spectrum use is concerned, regulators must also be willing to accommodate entrepreneurial ideas and be prepared to move ahead even of the venture capital community. If their natural risk aversion is unchecked it can easily become excessive and detrimental to the country. To avoid this outcome, the NRA should have within its walls some board and staff members willing to advocate innovation, to suggest new ideas and to pioneer them in the face of opposition from more conservative stakeholders. These are rare people willing to sacrifice a "quiet life" and risk failure for the sake of ideas, and regulators should generally do more to find and nurture such individuals.

Getting the right balance of bureaucracy versus innovation and careful management of a critical national resource versus enabling new applications is perhaps the most important of all challenges for a regulator.

1.4   International spectrum regulation

It is often noted that radio waves do not respect national boundaries, and that in consequence international co-ordination of spectrum use is needed to prevent mutually destructive interference – especially as most countries have land borders and are not remote islands.

---

[7] UWB was a technology requiring access to large amounts of spectrum at low power levels. The actual power levels that might be used without causing interference were a subject of much controversy and the regulatory community took a decade or so to address these and pass rules enabling UWB. By that stage, all of the start-up companies that had been pioneering its introduction had run out of funding and were no longer in existence.

The governance of spectrum use on a global basis is a core responsibility of the International Telecommunication Union (ITU), a body with a membership of nearly 200 countries. The ITU is a specialized agency of the United Nations with its headquarters located in Geneva, Switzerland. The ITU is not a global regulatory authority in the way that a national regulator is within its own jurisdiction, since the rules for international regulation and co-operation are written by those governed by them, i.e. by the member states of the ITU. These rules are administered by the ITU's Radiocommunication Bureau (BR), and conformity with them is based on goodwill rather than on the kind of regulatory sanctions found at the national level. The mission of the ITU in the spectrum management sector is, inter alia, to ensure rational, equitable, efficient and economical use of the radio frequency spectrum by all radiocommunication services, including those using satellite orbits, and to carry out studies and adopt recommendations on radiocommunication matters.

ITU World Radio Conferences are held every four years. One of the main tasks of the conferences is to review and, if necessary, revise the Radio Regulations, the international treaty governing the use of radio frequency spectrum and geostationary and non-geostationary satellite orbits. Typically a band is assigned to one or more co-primary uses and also to secondary uses. Primary uses are protected from interference from secondary uses, but not vice versa. These assignments may be different in each of the three ITU "regions":

- Region 1 comprises Europe, Africa, the Middle East west of the Persian Gulf including Iraq, the former Soviet Union, and Mongolia;
- Region 2 covers the Americas, Greenland, and some of the eastern Pacific Islands; and
- Region 3 contains most of non-former-Soviet Union Asia, east of and including Iran, and most of Oceania.

The most important decisions are taken at World Radio Conferences (WRC); more detailed decisions are taken at Regional Radio Conferences covering the three regions: such regional conferences cannot modify the Radio Regulations, unless approved by a WRC, and the Final Acts of the conference are only binding on those countries that are party to the agreement. Then, consistent with the Radio Regulations, individual countries may impose additional constraints on spectrum use, which are often recorded in a frequency allocation table.

Major decisions taken at recent World or Regional Radio Conferences include the following.

- At a Regional Conference in Geneva in 2004, terrestrial broadcasting frequencies were replanned to provide for digital terrestrial television. These decisions superseded earlier analogue plans agreed at a Stockholm conference in 1961.
- At the World Radio Conference held in Geneva in 2012 agreement was reached that the 700 MHz band, previously allocated to broadcasting, would be changed to mobile use in a number of regions in the world.

This last example is of interest since the decision was taken more quickly and with less notice than is the ITU norm. Typically, the full cycle of an item emerging, making its way up the agenda and being decided at a conference has taken up to a decade. Yet technological advance by spectrum-using firms often moves at a far faster pace. This observation has led to suggestions that the ITU process needs quickening across the board.

A further significant regional actor in spectrum management is the European Union (EU), the member states of which now routinely agree a common policy for ITU conferences. At present, the three main goals of the EU's Radio Spectrum Policy are to harmonize spectrum access conditions to enable interoperability and economies of scale for wireless equipment; to work towards a more efficient use of spectrum; and to improve the level of information about the availability of spectrum, its current use, and future plans for its use. Accordingly, for the first time in 2012, the European institutions (the Parliament, the Council of Ministers, and the European Commission) jointly produced a Radio Spectrum Policy Programme (RSPP). This defines a roadmap for the next steps in spectrum policy. It focusses on the spectrum needs for high-speed (4G) wireless broadband systems as a key action under the Digital Agenda for Europe. It also takes account of the requirements of other areas (such as audiovisual, transport, research, environment protection, or energy), while safeguarding essential defense, emergency, or earth observation requirements.

In 2013 proposals surfaced within Europe for a more vigorous attempt to get away from separate mobile markets in each country in favor of a single market providing mobile service across many member states. In the European Commission's 2013 proposal, this would have required changes to the spectrum assignment process within the EU, including the synchronization of auctions.

The list of other international organizations involved in some way or another in spectrum management is almost endless. It includes military alliances, such as NATO, which co-ordinate members' military spectrum use, and interest or stakeholder groups, such as radio astronomers, which seek to preserve and, if possible, extend their own access to spectrum.

In terms of the matters discussed in this book, the key issue is the extent to which the international arena is a major one affecting the methods of spectrum management employed and thus the degree to which they lead to more or less efficient outcomes. The answer to this question depends both on the country where the spectrum is used and on the nature of that use. Clearly a large country like Australia with no close neighbors will be much less constrained by international considerations than a small landlocked country like Luxembourg. Equally a non-geostationary satellite service will inevitably be international in its scope and regulation, whereas a communication system used by a small taxi firm will be local. In summary, as our subject matter unfolds, it will become clear that international constraints are not excessively constricting in many areas, and that spectrum management decisions taken at national level, or at regional level within a federal system such as the European Union, have a profound effect on, for example, the level of use of market instruments and the forms of spectrum sharing adopted.

1.5 Differences across countries and regions

Every country in the world is endowed, democratically, with spectrum which can be allocated and assigned within the same frequency table. However, the use made of it varies hugely. This is illustrated by the data collected by the UN Broadband Commission on mobile broadband penetration in many different countries [9].

At the top of the list are countries with a penetration of over 70 per 100 inhabitants (some of them in excess of 100). These include the leading countries in Asia (Singapore, Japan, Korea, Hong Kong etc.), in Northern Europe (Finland, Sweden, etc.), and the United States. At the bottom are many countries in Africa and some in Asia where the effective penetration rate is zero.

The data on penetration of mobile voice in developing countries are much more encouraging, and a growing number of studies show that access to this communications technology benefits the economy in general and certain groups of producers in particular.[8] Mobile data services are likely to have similar effects in the near future.

Are such differences across countries and regions likely to require completely different spectrum management regimes? It is clear that, in countries without fixed communications networks, mobile or wireless services are even more crucial for the development of the sector and for realizing its economic benefits than in those countries with good infrastructure. At the same time, developing countries may face less acute spectrum shortages than countries much further down the diffusion path for mobile data, but they are as likely as developed countries to have assigned a generously large share of spectrum to public-sector uses, especially to defense.

A significant feature of the fiscal situation of developing countries also abuts developing countries' spectrum policy. Their governments' capacity to raise taxes or other revenues is often limited. Mobile operators can, in such circumstances, be an attractive taxation target, and in some countries up to 40% of the cost of ownership of a mobile phone comprises various government taxes and tariffs [10]. More directly, proceeds from spectrum auctions can be a significant form of government revenue, making it tempting to restrict the supply of spectrum in order to enhance revenues. But this goal, from which no government is entirely immune, is accomplished at the cost of making available to the public less competitive, more expensive, and inferior mobile services, and at the cost of significantly reducing the tax base in the future.

Taking these factors into account, there is no clear reason why certain key attributes of the spectrum regimes of developing and developed countries should not be the same. In particular, spectrum should be made available for use in the marketplace for services, either in licensed form or as a commons. It should be recognized that in an auction, if competitive conditions in the market for spectrum and in the market for services are favorable, low spectrum prices should feed into low service prices, which benefit the economy. The regime should promote the long-term goals both of having an efficient allocation of spectrum among existing services, and of promoting new ones. Disciplined and efficient use of spectrum by the public sector is highly desirable. Finally, a spectrum

---

[8] See Chapter 13 below.

regulator in a developing country need not be an innovator in spectrum management, but can prepare itself to be a "fast follower" of successful new approaches tested elsewhere, and "leap-frog" less successful approaches.

1.6 Global, regional, or national spectrum management?

We have described above a complex multilevel system of spectrum regulation, with authority distributed at different geographical levels. In the world of regulation, this is anything but unusual [11], but we can nonetheless try to identify certain principles which might underlie an efficient arrangement.

A major one of these has to do with interference. As discussed in more detail in Chapter 2 below, signals transmitted by spectrum users in adjoining geographies (or bands) can interfere with one another, making some or all of them wholly unintelligible. Avoiding this outcome requires some co-ordination. This can be accomplished in a number of ways, which are discussed further below. But interference of this kind is an important species of spillover or externality, which must be taken into account in determining where to locate regulatory authority.

The implications of this for spectrum regulation vary with circumstances. Thus use of private mobile radio by a local taxi firm is in most cases unlikely to interfere with a similar use in another country. Terrestrial broadcasting, by contrast, is a high-powered activity which easily crosses national boundaries. In Europe this necessitated complex international agreements made first in Stockholm in 1961 to co-ordinate analogue terrestrial broadcasting, replaced by a later agreement made in Geneva in 2006 to deal with digital terrestrial broadcasting. In the case of satellite communications, even regional co-ordination is not enough; a global approach is needed to deal with it.

It follows from this that the need for higher-level spectrum management varies with the coverage area of the relevant technology. But while there is some link to frequency, with higher frequencies often resulting in small coverage zones, this is not always so; for example, satellites are at bands above 10 GHz where coverage for terrestrial systems is considered typically small. Some bands can be used for services with small coverage but then repurposed to services with larger coverage, or vice versa. Hence, while there is clearly scope to differentiate level of management according to service, this would prove complex to achieve in practice unless a very simple approach is adopted, such as the satellite/non-satellite split used today.

A second key consideration has to do with economies or diseconomies of scale. These operate at various levels. It may be more efficient for countries to share certain spectrum management activities, such as technical studies or international representation, and this may lead to the formation of a supra-national spectrum regulator. For example, many studies are performed on behalf of all European regulators by the CEPT, saving the national regulators the cost of conducting each study independently.

More importantly, co-ordinated decisions may lead to more efficient outcomes in downstream markets where spectrum is used. Thus it may facilitate entry and the development of new services over a wider area if spectrum licensing is performed centrally or even synchronized. Some of the success of the US in leading in wireless innovation can be ascribed to its "single market" covering over 300 million inhabitants.

Harmonization of frequencies for particular activities may also permit economies of scale in the production of transmission and reception equipment, notably devices, and thus generate cost savings which permit user price reductions. However, this consideration loses some of its force in cases where a service is already widely diffused. Thus there may be about five billion or so SIM cards in use in the world, the bulk of them housed in various mobile devices, most of which have been made capable of multiband operation at a very small incremental cost. Efficient scales of production are thus likely to have been realized in a number of bands. Moreover, operators can reasonably be relied upon to avoid seeking to provide service on a band which is not in wide use, unless that band has compensating advantages. Use of bands supplied only with costly equipment is likely to be discouraged by the self-policing of competitive operators.

Multinational spectrum management becomes more important for smaller countries and particularly those with multiple borders. Some central European countries have up to nine neighbors, resulting in a significant need to co-ordinate activities and the potential for much unused spectrum set aside for guard zones in border areas. By comparison, Australia has very little in the way of borders and is well isolated from most other countries such that it can develop policy with little regard for its neighbors. The US has two major borders (Mexico and Canada), but these fall some distance from most populated areas and the size of any guard zones will be a tiny percentage of the overall landmass. Hence we might expect greater benefit from multi-country management in Europe than in Australia or the US.

The question of innovation is an important one. Arguably, the US is a consolidated spectrum market of approximately the population and size of Europe. The US has been highly innovative in introducing new spectrum management techniques and has been the first, for example, to allow UWB and TV white space, and is pioneering new auction techniques. This suggests that innovation is not inhibited by size, but more by culture and regulatory structure. Thus there are concerns that European Commission-led spectrum management could become unwieldy and overly bureaucratic, stifling the kind of innovation that Ofcom have exhibited in the UK but few other European countries have matched. Instead, as noted below, the process of loose co-ordination through the CEPT and via ad hoc working groups as needed[9] continues.

This suggests that:

1. Countries which are large geographically, have few borders, and have a large internal market will gain little from multinational spectrum management and should continue to manage nationally. This category includes the US and probably China and India.
2. Countries which are large geographically and have few borders but have a small internal market will gain from international management, but this can be restricted to specific services and can probably be achieved using existing regional and global spectrum organizations. Australia is a good example of such a country.

---

[9] For example, the Geneva 2006 group, set up to co-ordinate the use of UHF TV transmitters across Europe after the digital switch-over.

3. Countries which are small geographically and/or have multiple borders and have a small internal market will gain significantly from regional spectrum management. While this could be for certain bands or services only, such an approach is complex to define and moving directly to multinational management may be preferable.

However, while in principle we see benefit for those regulators in the third category to form a regional spectrum manager, we see practical obstacles to such an approach that currently appear intractable. There are strong forces preserving the status quo and only weak voices advocating the advantages of regional management.

While these questions are discussed everywhere, they have been debated with particular intensity in Europe, partly because European nations are generally small and crammed closely together, and partly because of the wider tensions within the 28-nation European Union over the appropriate degree of federalism.

Europe is widely credited with benefiting from the success of the passage in 1987 of the GSM Directive, which was influential in ensuring the success of GSM and enhancing the strengths of the European telecommunications industry. However, since then there have been more failures than successes, with directives such as that for ERMES paging occupying frequency bands but not resulting in any service. More recently the Ultra-wideband (UWB) Directive was seen as too restrictive and as stalling the development of equipment worldwide. As a result, there have been few binding directives in recent years – for example, none on TV white-space spectrum, which would appear to be an area ripe for pan-European activity.

The EC has issued recommendations to make licensing and allocations technology-neutral, and to ensure open competition and free movement of equipment. These have been developed and delivered through EC-established bodies such as the Radio Spectrum Policy Group (RSPG) and the Radio Spectrum Committee (RSC). However, national regulators tend to adhere to these recommendations only when it suits them, and there is little indication that the work of the EC has resulted in more pan-European harmonization than might have been delivered through CEPT. The relationship between the EC and CEPT has also been somewhat fraught, with the EC wishing to make CEPT its technical arm while CEPT (which includes many members which are not member states of the European Union) has wished to maintain its autonomy.

In 2013, the European Commission (EC), the executive arm of the EU, brought forward the so-called "connected continent" proposals, which sought to establish a single digital market and included proposals for expanded powers for the Commission to co-ordinate spectrum awards [12]. By 2015, however, many of the proposals, including those on spectrum, had lapsed, although they may be revived.

The situation has, arguably, resulted in even more delay and uncertainty than would have been the case had the EC not existed. National regulators often feel that they should not move ahead with new ideas until it is clear whether the EC will make an enactment which may cause them to modify or even retract regulations they might have made. EU decision-making can sometimes proceed at the pace of the slowest regulator, stalling innovation.

Europe has lost much of its technological leadership in radio systems over the last 20 years. The extent to which this is caused by regulation is hard to determine, but it is telling that the most powerful companies are now those in the US, where regulation is not federal in nature.

A pan-European regulator, if set up with appropriate powers and independence, and if staffed by those with an inclination toward innovation rather than public service, ought to be able to deliver a more innovative outcome across Europe as a whole than occurs today. However, as noted above, the current structures and approaches at a European level tend to limit, rather than facilitate, innovation.

1.7 Successes and challenges

Spectrum management has been practiced for over 100 years – the first Act of Parliament in the UK was in 1903, for example. But despite this, it is still very much a work in progress. This is because of aspects such as changing technology, growing demand for the spectrum, and changing global demographics. We conclude this introductory chapter by looking at what has worked well to date, and the problems and challenges that remain.

Aspects that have worked well include the following.

- **Providing stability.** The slowly changing nature of regulation, with its consultative approach, and the long timescales on spectrum licenses have provided the guarantees that commercial organizations have needed to commit to the rollout of expensive national networks which often have payback periods of a decade or longer.
- **Freedom from interference.** Serious interference is rarely experienced by users of spectrum. Where interference does occur it often tends to be deliberate – for example the deployment of pirate radio stations. Otherwise, the risk-averse spectrum planning laws tend to result in a very low probability of any interference occurring.[10]
- **Working in an open and consultative manner.** Regulators are generally seen as working in a manner that is very open and transparent. This can help all to understand the decisions being made and help avoid unexpected problems.
- **Auctions as a means of assigning spectrum.** Auctions have been increasingly used since the late 1980s as a means of assigning scarce spectrum. While there has been controversy over the amounts raised and occasional failures in terms of unsold licenses or uncollected fees, they have generally worked well in terms of transparency and initial efficiency.

But there have been many problems and challenges, including:

- **Making rapid changes of use.** The downside of a stable and consultative regime is that it cannot react quickly to changing demands. This has been apparent with the mobile data explosion, requiring change on timescales of one to two years rather than

---

[10] Some might argue that this is more of a failure than a success, and that a better regime would result in some limited number of interference cases as it sought the optimal point between maximizing usage of the spectrum while minimizing interference.

one to two decades. This can prevent the emergence of desirable services, lead to inefficient solutions by those deploying networks, and make it difficult for start-ups and entrepreneurs to develop solutions needing access to the radio spectrum [13].
- **Better enabling market mechanisms.** One of the solutions to a slow speed of reaction would be to move decisions to the market, where solutions such as trading can enable rapid change of ownership and usage. However, these mechanisms have not been widely adopted since their introduction a decade or more ago. There are many possible reasons for this, but regulators could do more to facilitate international markets, improve liquidity and clearly signal the end to the "clear and auction" approach adopted today.
- **Being innovative.** Innovation is rightly seen as important in driving new ideas and improvements in productivity. As well as innovations in the services that use spectrum, there are many possible innovations in spectrum management, such as better shared usage of spectrum, dynamic spectrum allocation, and more advanced modeling tools. However, regulators have little incentive to explore and introduce these and some disincentives in that any new approach will involve some risk. As a result, the speed of innovation in regulation has, with some exceptions, been slow and does not appear to be meeting the demand caused by service innovation.
- **Finding the right balance between national and international regulation.** As discussed above, spectrum legislation tends to occur at a national level. However, radio waves propagate freely across national borders and there are ever-increasing needs for international harmonization of spectrum use to enable economies of scale and allow roaming of devices across borders. Also spectrum markets such as trading would likely work better if there were an international market for spectrum, allowing users to readily build spectrum holdings across a region rather than just a country. However, regulators have tended to resist any attempts to centralize spectrum management powers in bodies such as the European Commission. While there are valid concerns with abrogation of national sovereignty, regulators could have been more proactive in seeking a well-reasoned position.
- **Better balancing the needs of the private and public sectors.** The public sector has historically readily gained access to spectrum but been reluctant to relinquish this as the needs of the private sector have grown. Regulators have tried to redress this but found it difficult to do so for a range of reasons, such as a lack of power to force certain behaviors and unwillingness to expose the public sector to the full force of market access. This remains a problem without a clear solution.
- **Balancing competition concerns with radio spectrum usage.** There is tension between competition and spectrum efficiency. The most efficient use of spectrum from a technical viewpoint would be for there to be a single network for any given service such as cellular telephony. However, having competing networks has been shown to spur innovation, reduce prices, and increase consumer choice and quality. Finding the right balance between spectrum efficiency and competition is challenging and can be a moving target as the economics of service provision change.
- **Working across multiple sectors to optimize spectrum use.** Increasingly, networks are complex entities comprising fixed and mobile components. Spectrum can be used

more efficiently in small cells, but for these to be cost-effective there is a need for widely available low-cost backhaul. Regulators that manage both fixed and mobile networks, such as Ofcom in the UK, could look to require fixed operators to provide better backhaul connectivity as part of a more holistic approach to maximizing the economic value of telecommunications. However, at present regulation tends to occur in silos and on a more backward-looking basis.

It is clear that many of the approaches that have been working well have a downside in causing problems in other areas. Spectrum management tends to be a balance between many competing requirements and a perfect solution may not be possible. Instead, regulators need to strive to find the best balance between competing demands and to review this as demands change over time.

References

[1] Ofcom, "Spectrum Attribution Metrics" (December 2013).
[2] R. Thanki, "The Economic Significance of Licence-Exempt Spectrum to the Future of the Internet" (March 2012).
[3] See FCC, "Auctions Summary," at http://wireless.fcc.gov/auctions/default.htm?job=auctions_all.
[4] See www.legislation.gov.uk/ukpga/2003/21/contents for the Communications Act.
[5] See www.erodocdb.dk/docs/doc98/official/pdf/CEPTRep031.pdf.
[6] See Ofcom, "Digital Dividend: Clearing the 800 MHz Band," at http://stakeholders.ofcom.org.uk/binaries/consultations/800mhz/statement/clearing.pdf.
[7] See ITU definition at http://life.itu.int/radioclub/rr/art01.htm.
[8] See http://consumers.ofcom.org.uk/tv-radio/radio/tackling-pirate-radio.
[9] Broadband Commission, "The State of Broadband 2014: Broadband for All," ITU/UNESCO, pp. 98–99.
[10] M. Cave, "How Strong Is the Case for the Fiscal Exceptionalism of the Telecommunications Sector?" (2012) 2 *International Journal of Management and Network Economics* 322.
[11] R. Baldwin, M. Cave, and M. Lodge, *Understanding Regulation*, Oxford University Press, 2011, Part 5.
[12] See http://ec.europa.eu/digital-agenda/en/connected-continent-single-telecom-market-growth-jobs.
[13] J. Hausman, "Valuing the Effect of Regulation on New Services in Telecommunications" (1997), *Brookings Papers on Economic Activities, Microeconomics*, pp. 1–37.

2 The technical challenge

2.1 Introduction

The most fundamental reason for managing radio spectrum is to avoid interference between users, or at least to ensure interference is manageable and optimizes the capacity of the radio spectrum. This chapter provides the background material necessary to understand how interference occurs and the mechanisms to mitigate its effects. This is fundamental to the understanding of the spectrum management discussions throughout the rest of this book.

The chapter broadly describes the characteristics of a radio signal, how it travels from a transmitter to a receiver, the ways that interference can occur in the receiver, and how these can be avoided using technology.

2.2 Transmitting a radio signal

2.2.1 Defining a radio signal

A radio signal is the emission of an electromagnetic wave from an antenna. Electromagnetic waves cover a broad range of frequencies, from radio waves up through visible light and beyond into X-rays. Radio waves are generally considered to be those where the frequencies range from around 10 kHz up to 300 GHz. They are formed when an electrical signal passes into any conductor that has the ability to radiate signals. Antennas are designed for this purpose, but cabling can be an unwanted radiator if not properly shielded, and even semiconductor chips tend to radiate unwanted signals.

2.2.2 Components of a signal: power, bandwidth, and mask

At its simplest a radio signal has three key components – power, bandwidth and frequency. The power is the amount of energy radiated from the antenna, typically measured in watts. The bandwidth is the width of the different frequency components of the signal, measured in hertz, and the center frequency is the frequency of the transmitter when not transmitting any information (said to be unmodulated), also measured in hertz. For example, a GSM handset might transmit 1 W of power at a center frequency of 900 MHz with a bandwidth of 200 kHz.

As might be expected, there are some subtleties in each of these parameters. The antenna used by a device can radiate energy equally in all directions (an "omnidirectional antenna") or it can focus energy in one particular direction (a "directional antenna"). The former is like a light bulb in a room, the latter like a torch. The effect of a directional antenna is to increase the energy received in the direction of the beam and decrease it elsewhere. In order to avoid a transmitter being given a license for a particular power and then using a directional antenna to boost that power in a particular direction, transmitter powers are often specified as equivalent isotropic radiated power (EIRP). This is the power that would have been received had an omnidirectional or isotropic antenna been deployed. So if a directional antenna with a gain of a factor of ten (often described as 10 dB, as set out below) is being used and the license allows for a 1 W EIRP transmission, then the radio transmitter can only pass deliver 100 mW (which is 1 W less 10 dB) into the antenna.

The next complexity is bandwidth. It is simplest to think of a radio signal as the rectangular block shown in Figure 2-1. The block shows an idealized transmission where the radio signal maintains a constant power level across its entire allowed bandwidth and falls immediately to zero power outside this. In this case, measuring the bandwidth would be easy. In practice, it is not possible to generate signals of this form – it would require infinitely complex filters. Radio transmissions tend to have a structure more akin to that in Figure 2-2.

Here the upper line represents the emissions allowed by the regulator. Note that it is not rectangular but allows some emissions outside a central region, recognizing that this

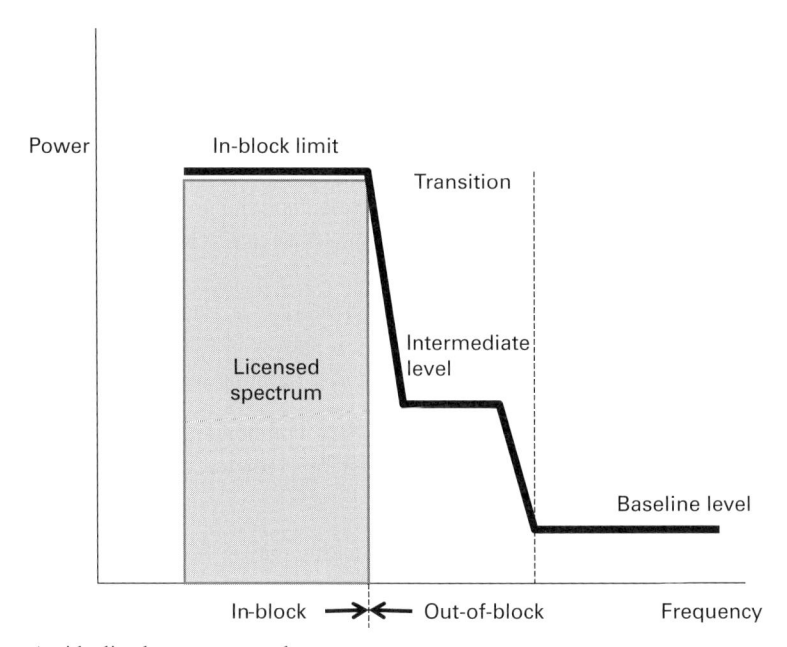

Figure 2-1. An idealized spectrum mask.

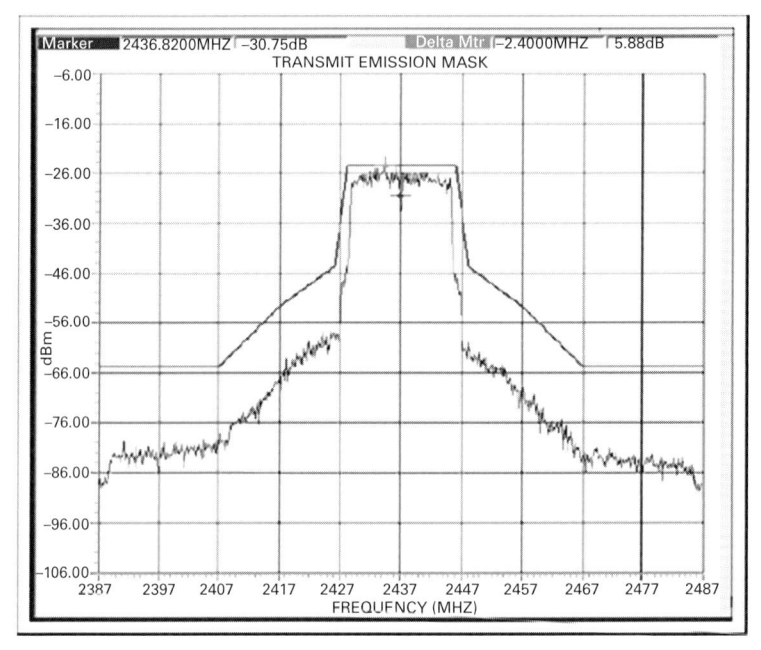

Figure 2-2. Actual transmitter emissions (lower line) sitting below an allowed mask (upper line).

will always be so in practice. The lower line represents an actual device which can be seen to keep its emissions within the limit.

A decibel, or dB, is a logarithmic measure of power where the relative power in dB is:

$$\text{power (dB)} = 10\log_{10}(P_1/P_0)$$

As a consequence, −3 dB is half-power, −6 dB is quarter-power and −10 dB is one-tenth of the power.

The bandwidth of such a signal is typically taken at the 3 dB point below maximum power. In this case the maximum power is −26 dBm (a dBm is 1,000th of a dB, a measure of power relative to 1 milliwatt, so $P_0 = 1$ mW) and the point at which the signal falls to −29 dBm essentially the width of the central section – in this case a little less than 20 MHz.

The upper line is known as a spectrum mask – a line representing the allowed emission levels versus frequency. This is also shown in Figure 2-1 where it is divided into an in-block level, a transition level, and a baseline level. Typically, the in-block element would correspond to the spectrum considered licensed to that user, but the transition and baseline levels might be in spectrum owned by others. As a result, the emissions in these areas need to be as small as possible commensurate with economically viable devices.

The combination of power and bandwidth define how much information can be transmitted by the signal. At its most fundamental this is described by Shannon's law [1], which states that the channel capacity C, meaning the theoretical upper bound on the information rate that can be sent with a given average signal power S through a communication channel subject to additive white Gaussian noise of power N, is:

$$C = B * \log_2(1 + S/N)$$

Where:

- C is the channel capacity in bits per second;
- B is the bandwidth of the channel in hertz;
- S is the average received signal power over the bandwidth, measured in watts;
- N is the average noise or interference power over the bandwidth, measured in watts; and
- S/N is the signal-to-noise ratio (SNR) of the communication signal to the Gaussian noise interference expressed as a linear power ratio.

So, broadly, the information carried rises linearly with bandwidth and as the logarithm of signal power. For many systems power is limited, either by regulation or by the practicalities of generating high-power signals economically, and hence bandwidth is the key variable. This is the reason why bandwidths have been getting progressively higher in mobile phone systems as higher data rates are desired. However, there is only a fixed amount of bandwidth available – below 1 GHz of radio spectrum there is, by definition, only 1 GHz (1,000 MHz) of bandwidth and hence pressure to use it from multiple different systems.

It is important to point out that Shannon's law applies on a "per-channel" basis. There are a number of techniques which rely on the creation of multiple channels (for example, by bouncing signals off walls) which means that one piece of spectrum can concurrently support more than one Shannon channel.

2.2.3 Problems: intermodulation, spurious emissions, and harmonics

As well as spreading the transmitted signal across a wide bandwidth, as shown above, radio devices can generate narrow but powerful emissions in frequencies some distance away from their center frequency. These are generally known as "spurious emissions" and can plague radio designers who have to go to some lengths to avoid their generation. There are broadly two sorts – harmonics and intermodulation.

Harmonics are signals that occur at multiples of wanted frequencies. The strongest are at low-integer multiples – so the second harmonic would be at twice the carrier frequency. For example, a GSM transmitter working at 900 MHz would also emit a signal at 1,800 MHz unless it was minimized through careful design. The reason for this is that it is very difficult to generate a perfect sine wave at the carrier frequency. Any distortions to the wave will effectively be the imposition of a higher-frequency sine wave at some multiple of the carrier (since all signals can be represented as a combination of sine waves). Second and third harmonics are particularly troublesome. On the plus side, they can be much reduced by filters placed at the output of the transmitter since they are some large distance in frequency from the wanted signal.

More problematic is intermodulation. This occurs when two different signals are mixed together. If these were at frequencies f_1 and f_2, then the mixing process can lead to signals at $(2f_1 - f_2)$ and $(2f_2 - 2f_1)$. If $f_1 + f_2$ are close together in frequency then the intermodulated signals can also be very close to the wanted frequencies. This makes them both difficult to remove with filters and particularly troublesome.

Intermodulation often occurs due to the combination of two separate transmissions. For example, two different cellular operators might use the same tower to site their antennas. If their transmitted signals are mixed together then intermodulation occurs. This can happen inadvertently in badly fitting bolts in the tower or it can happen in the receiving equipment itself (covered in section 2.4.1).

In general, with good radio design, spurious emissions are less of a problem than the wanted radio signal and the unwanted emissions on either side of it.

2.3 How signals propagate

2.3.1 Propagation in free space

Once a radio signal has been generated it travels outwards from the antenna in a manner similar to ripples on a pond [2]. However, since the signal can travel in all three dimensions, in the absence of anything to prevent it (so-called "free space") it actually spreads out as the surface of a sphere. The surface area of a sphere is proportional to the square of the radius and all of the power is equally distributed across the surface. Hence the received power falls away as the square of the distance from the transmitter. This is exactly what happens with light from the Sun, with power decreasing per square meter of "receiver" as the square of the distance. It never reaches zero, although does get to a point where it is undetectably small.

Free space is an ideal that is never realized in radio systems. First, antennas are not perfect radiators in all directions. Indeed, a typical dipole antenna radiates in a pattern more like a doughnut, equally in all directions around the antenna but very little in the upward or downward plane. Second, all sorts of things get in the way, from the ground to buildings. Finally, signals can be absorbed by the atmosphere. This tends to mean that the received signal is less than it would be in free space – so the free-space prediction is an upper bound on signal level. There are a few isolated examples of where this is not true, for example when a signal travels inside a metallic tunnel and so is constrained in a particular direction, but these are rare.

2.3.2 Different frequencies have different characteristics

In practice, the amount of signal received is also dependent on the center frequency of transmission. There are two reasons for this – antenna size and absorption.

The optimal size for an antenna is related to the wavelength of the radio signal – in fact it is a half-wavelength. At this point more of the radio signal is converted to electrical energy than for any other antenna. In turn the wavelength is inversely related to the frequency by the formula

How signals propagate

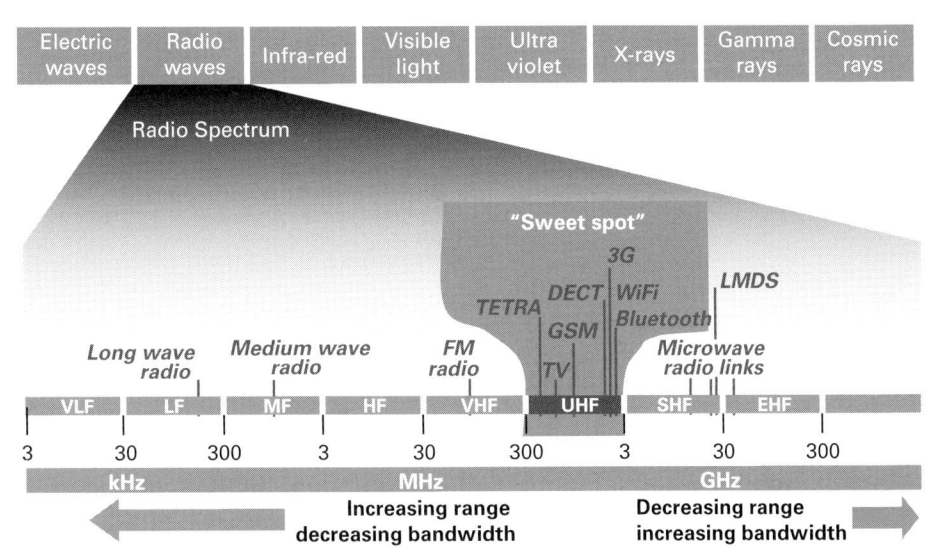

Figure 2-3. Radio spectrum and the sweet spot. Source: [3].

$$V = f * \lambda$$

Where V is the velocity of light, f the frequency and λ the wavelength. So a higher-frequency signal will have a shorter wavelength and hence a smaller optimal antenna. But the amount of energy in a radio wave is proportional to the percentage of the "wavefront" incident on an antenna. If an antenna is half the size it will see half as much wavefront and hence half the signal. So optimal antennas see decreasingly less signal as the frequency increases. This can be overcome by combining the output of multiple small antennas, but this is expensive, requiring multiple radio receivers.

Further, the higher the frequency the more a signal tends to be absorbed by the atmosphere and materials and the less it will diffract around obstacles. So in the real world, higher-frequency signals travel less far.

As a result, lower frequencies are preferred because they travel further. However, there is less bandwidth available at lower frequencies (there is 1,000 MHz below 1 GHz but 9,000 MHz between 1 GHz and 10 GHz). Most radio transmissions require as much bandwidth as possible. The best place to be is often the middle ground where propagation is still fairly good and there is significant spectrum available. This "sweet spot" is often considered to be approximately 300 MHz–3 GHz. The bottom limit of 300 MHz is the point at which the size of antennas (based on the half-wavelength described above) is too large for the antenna to be integrated into a portable device). This sweet spot is shown graphically in Figure 2-3.

2.3.3 Propagation in real life: the Hata model

Real-life signal propagation is much more complex – so much so that accurate prediction over more than a few hundred meters is nearly impossible. Signals interact with the ground, reflect off objects, pass through others, are absorbed by some, refract around others, and arrive at their destination via multiple paths.

For radio design purposes propagation is often thought of as consisting of three elements:

- The average signal level in a particular area (often a square of size around 50 m × 50 m).
- The variation in signal due to obstacles such as buildings, which is typically known as "slow fading."[1]
- The variation due to multiple signals arriving at the destination and sometimes canceling each other out, other times reinforcing each other, typically known as "fast fading."[2]

Neither slow fading nor fast fading can typically be modeled accurately but their distributions are known from multiple measurements.[3] So a radio system will normally have a minimum signal level for successful operation, which will allow a margin for slow and fast fading.

For the purposes of spectrum management the average signal level is generally sufficient, although there are exceptions explored further below. The average signal level is often estimated based on empirical models which have been developed from multiple measurement campaigns. One of the most widely used of these is the "Hata model," named after an eponymous Japanese engineer [4].

The Hata model for urban areas is formulated as following:

$$L_U = 69.55 + 26.1\log_{10}f - 138.2\log_{10}h_B - C + [44.9 - 6.55\log_{10}h_B] * \log_{10}d$$

For a small or medium sized city:

$$C_H = 0.8 + (1.1\log_{10}f - 0.7) * h_M - 1.56\log_{10}f$$

Where

L_U = path loss in urban areas. Unit: decibel (dB).
h_B = height of base station antenna. Unit: meter (m).
h_M = height of mobile station antenna. Unit: meter (m).
f = frequency of transmission. Unit: megahertz (MHz).
C_H = antenna height correction factor.
d = distance between the base and mobile stations. Unit: kilometer (km).

Other correction factors are available for cities and suburban environments of different densities.

---

[1] "Slow" because a mobile user, perhaps walking around, would take some seconds to pass through the shadow of a building, which is a long time in radio system operation, so the changes are slow relative to the adaptation of the system.
[2] "Fast" because the distance between these fades is around a wavelength, which is typically 10–30 cm at the frequencies used for mobile transmissions. Users can move this far in much less than a second.
[3] Slow fading is typically modeled as a log-Normal distribution with a zero mean and a standard deviation of around 6 dB. Fast fading is typically modeled as a Rayleigh distribution.

Taking each of the elements of the equation in turn, we first have a fixed part (69.55) which relates to the expected path loss at 1 km from the antenna (the model effectively works at distances above 1 km from the transmitter). Next we have a component related to the frequency that reflects the increasing propagation loss with increased frequency, as explained earlier. Then comes a factor based on the height of the base station, recognizing that signals from higher towers travel further. Then there is a correction factor that allows for different densities of buildings. Finally there is an element related to the distance. If free-space path loss applied then this would be d^2, which in logarithmic terms is $20\log_{10}d$. In this case the term is related to the base station height – for 10 m the term would be 38.5 and for 100 m (a very tall cellphone tower) it would be 32. This is somewhere between d^3 and d^4, showing that signals propagate much less freely in the real world than in free space.

The Hata model is an approximation. If a spectrum manager wanted to know in general how far a signal from a licensed transmitter might travel then it would be a good starting point. However, a mobile operator deploying a network would use a propagation modeling package which considered each cellphone site, looking at the terrain around it, the type of "clutter" that the signal traveled through or over (e.g. buildings, forest, farmland, water), and the known parameters of transmitters and receivers.

2.3.4 Buildings

Buildings present many complexities to radio signals and to modeling. The extent to which a signal will travel into (or out of) a building is called the "building penetration loss" and depends on factors such as the material the building is made of, the size of windows and whether they are coated with metallic materials, and the angle of incidence of the incoming radio wave. Since this information is rarely known for each building then approximations are made. Building penetration loss is often modeled as a log-Normal distribution with an average of around 12 dB and a variation (σ) of around 5 dB. This means that on average the signal will be 12 dB weaker inside the building than outside, but that there is a 1% chance that it will be 27 dB weaker (the average plus 3σ).

Propagation within buildings can also be of interest for systems such as Wi-Fi. It is similarly difficult to predict, depending again on the construction material of interior walls and floors, the positioning of furniture (especially large metallic objects like filing cabinets), and even the extent to which signals can radiate through an outside window, reflect off a nearby building, and return via a different window. Broadly, it is modeled by considering the likely number of interior walls and floors the signal would need to pass through and assigning a loss per wall and per floor. The loss parameter could be generic for that type of building (e.g. office, residential) or could be based on actual measurements.

While the building penetration loss is problematic for cellular systems aiming to provide coverage inside buildings from transmitters based outside, it can be beneficial for spectrum management as it can provide something of a barrier between systems operating inside the building and those operating outside, allowing some degree of sharing of the radio spectrum.

2.3.5 Minimum coupling loss

A parameter that is sometimes of importance in spectrum management is the minimum coupling loss (MCL). This is the smallest propagation loss that can occur between a transmitter and a receiver. It is of importance when considering the possible interference from one system to another – for example from a mobile phone into a TV receiver when they are in adjacent frequency bands. The MCL is the lowest possible amount by which the mobile phone signal could be attenuated leading to the maximum interference. The spectrum manager could then choose to set the mobile phone power levels so that the resulting signal did not cause harmful interference.

There are problems with this approach that we will return to later, predominantly that it is conservative in that the MCL is rarely observed in practice so protecting the receiver to this level may be overly pessimistic. This reduces the possible use of the spectrum by other services and hence the overall utility that a country can enjoy from a scarce resource.

To calculate the MCL it is first necessary to define the geometry of the situation. For example, it could be decided that the mobile phone is in the same room as a TV receiver with a set-top aerial. They might be as close together as 1 m with no obstacles in between. Then the MCL would be the free-space loss at 1 m from the antenna (around 32 dB). Alternatively, it might be decided that the mobile phone is outside at street level and the TV receiver is using a rooftop antenna. Then the important characteristics are:

- the distance from the phone to the rooftop antenna,
- the gain of the rooftop antenna in the direction of the phone,
- any losses due to the antenna polarization of the phone and the rooftop antenna being different, and
- any losses in the cabling between the rooftop aerial and the TV receiver in the home.

Each of these can be determined and added to produce an MCL (around 52 dB in this case).

Calculation of the MCL is typically much simpler than a general propagation estimation because the two devices are often so close that free-space propagation can be considered to apply. However, defining representative geometries is much more difficult and can call for judgment. As the example above shows, there can easily be 20 dB difference between two interpretations of the likely interference cases. This is a huge difference relating to a range for the mobile phone of perhaps 100 m in the worst case and 3 km in the best case.

It is possible to adopt a probabilistic approach to MCL, allowing for a distribution of, say, distances of the mobile phone from the antenna. This leads to a distribution of MCLs and it might be appropriate to take the point where only 1% of cases cause a worse result (more interference) rather than the absolute worst case.

2.3.6 Weather and anomalies: ducting and seasonal variations

Even when a prediction is accurate, it may vary over time. Some variations are short-term, caused by factors like a bus briefly stopping between transmitter and receiver. Others are seasonal – for example trees with leaves absorb more radio waves than those without, resulting in higher propagation losses in the summer. Finally, at some frequencies and for some types of transmission an effect known as "ducting" can occur where signals traveling upwards can be contained in a layer of the atmosphere, bouncing off an upper discontinuity to a lower one, bouncing back up off that and so on until they finally emerge some distance away [5]. Ducting primarily occurs at the frequencies used for TV transmitters and with the high transmitter sites that TV often uses. This can result in TV signals traveling hundreds of kilometers in some cases and causing unexpected interference. Ducting tends to occur when there are large and stable high-pressure zones. As a result, broadcasters often allow a further margin in designing reception to take this into account.

2.4 Mechanisms of interference

2.4.1 Overview

Interference occurs when an unwanted radio signal combines with a wanted radio signal and reduces the quality of the wanted signal. There are varying degrees of interference, from mild, where the effects on the wanted signal are minor, through to complete interference, where the wanted signal can no longer be received. In spectrum management terms the latter is often referred to as "harmful interference," although there are no quantitative definitions as to exactly what level of interference is harmful.

It is a key point that interference only occurs in a receiver. Transmitters are unaffected by interference and even if there are multiple radio waves present, if there are no receivers then nobody suffers from interference.

There are multiple mechanisms that can cause interference. These are shown in Figure 2-4. In this figure there are three different users of spectrum. The top two both use the same spectrum but in different geographical areas. Any interference from one to another is termed "co-channel" interference. The bottom user is in the same geographical area as the top left user and these users are using different but adjacent frequencies F1 and F2. Because transmitters are imperfect and have some emissions in adjacent channels, the top user will generate some interference on frequency F2. This is often known as "adjacent channel leakage." Also, the receivers of the users operating on F2 will have imperfect filters and will not be able to remove the entire signal on F1. Some will end up within the receiver as interference. This is known as "adjacent channel selectivity." Finally, not shown on the chart, there can be interference-induced issues in a poorly designed receiver – for example the interference can overload amplifiers within the receiver, causing it to malfunction.

All of these mechanisms can occur at the same time. In particular, for users in the same area a receiver may have both adjacent channel leakage interference and adjacent

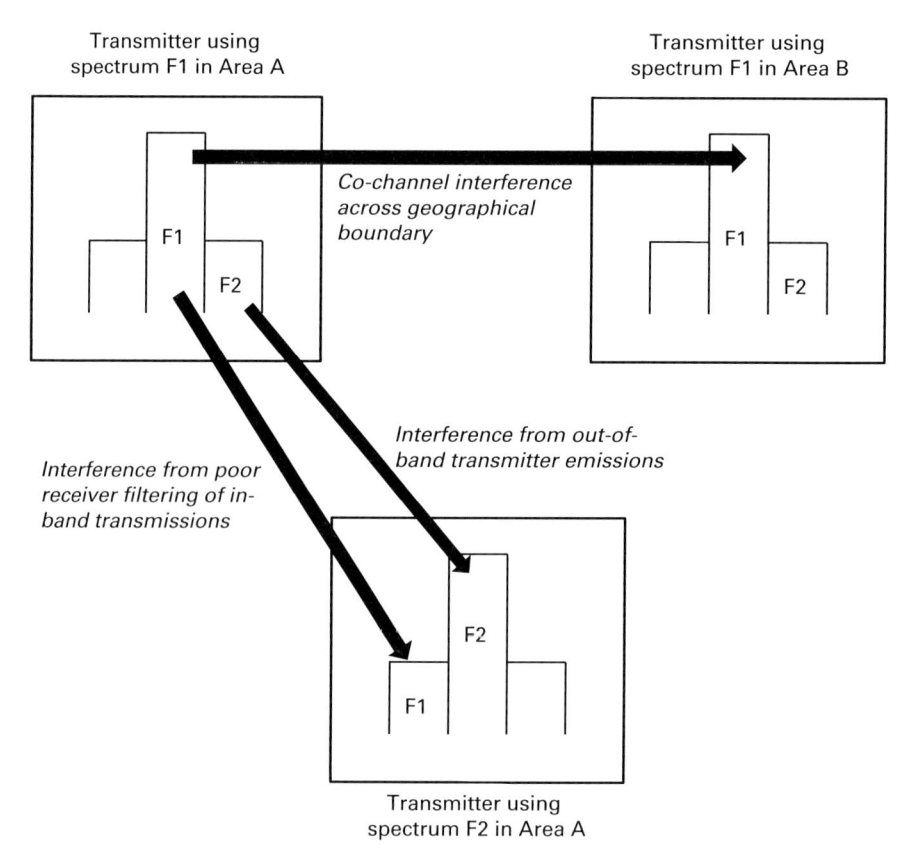

Figure 2-4. Mechanisms that can cause interference.

channel selectivity interference. These two effects add together to increase the overall interference.

Note that the objective of the spectrum manager is not to remove interference totally. This would be impossible as radio signals travel indefinitely. Indeed, in cellular systems the networks are designed to self-interfere to a carefully defined level. At this point the capacity of the network is optimized as relatively high power levels are allowed for transmission and the interference is such that while noticeable it does not materially degrade reception. The objective of the spectrum manager should be similar – to allow interference up to a level that is just below the threshold of being noticeable.

Each of the different types of interference is discussed in more detail below.

2.4.2 Co-channel interference

The strength of a radio signal reduces with distance from the transmitter. Once far enough away from a transmitter site the same frequencies can be reused by a different transmitter. The distance between them should be such that someone in the middle cannot receive a signal from either. This area in the middle is often termed a "guard zone." In an ideal world it would be very small, but in practice, given the difficulties in predicting propagation, the guard zone is often of some size so that the regulator can be sure that the signal from each transmitter has decayed away more than sufficiently by the time it reaches the coverage area of the other transmitter. Clearly there is a matter of judgment here between the desire to avoid interference and the need to make the maximum possible use of the radio spectrum.

In some cases the co-channel use is within a country. For example, regulators often license business radio use on a transmitter-by-transmitter basis. A taxi company in one part of the country can be given the same frequency as another taxi company in a different part of the country if far enough separated. Co-channel use can also occur across national boundaries. In this case, since radio signals do not stop at boundaries, the two national regulators need to work together to give out licenses that allow for guard zones around the boundary or other ways of license holders working together to avoid interference.

2.4.3 Adjacent channel leakage

Signals from a transmitter that spill out into adjacent channels are known as adjacent channel leakage (ACL), as shown in Figure 2-2. This signal is typically at a much lower level than the wanted transmission – often licenses will specify that the ACL is something like 40–50 dB less than the wanted signal. So if the allowed transmission is 1 W (30 dBm) then the adjacent channel signal will be required to be around 0.1 mW (−10 dBm) or less. This signal is then further attenuated by the propagation loss between transmitter and receiver, which in its smallest case will be the minimum coupling loss. This might be a further 40–50 dB. So overall the signal might be around 80–100 dB less than the transmit power allowed. However, radio devices can often decode signals at around 130 dB less than the transmitted level, so even this reduced signal level can lead to interference.

As with other types of interference, the regulator should allow some ACL. Not to do so would make the transmitter costs overly high. Often ACL levels are actually set within standards bodies – for example 3GPP sets the allowed out-of-band emission limits for 3G and 4G transmitters. However, such standards bodies will set these levels to be optimal for interference from one system to another of the same technology (e.g. from one 4G network to another 4G network in neighboring bands). This level may not be optimal when the neighbors use differing technologies. Generally, the regulatory response where this is the case is to insert a "guard band" – an unused piece of the radio spectrum which acts as a buffer between different users. As with a guard zone, such bands are undesirable because they reduce the efficiency with which the radio spectrum can be used.

The impact of adjacent channel leakage depends very much on the neighboring uses. For example, if both are cellular networks and both are frequency division duplex (FDD), then neighboring transmissions will both be from either base station or both be from terminals. Considering the base station case, a terminal from one network (say network A) is unlikely to be extremely close to a base station from network B because base stations tend to be in elevated positions. This means that the MCL is greater and the impact of B's ACL is reduced. Compare this to the case where both networks are using time division duplex (TDD), when a terminal from A might be transmitting on a neighboring frequency while a terminal from B is trying to decode a transmission from its base station. These two terminals could be very close together, e.g. in a crowded railway carriage, and so the MCL small. In this situation the interference impact will be much more significant. The fact that the impact of the interference (not its absolute level) depends on the technology and usage of the license holders makes technology-neutral spectrum management complicated – a topic that we will return to later, in section 10.2.

2.4.4 Adjacent channel selectivity

The antenna of a radio receiver will receive all the radio signals in the vicinity across the entire frequency band. However, the receiver will only be interested in the wanted signal. All other signals must be filtered out before decoding the wanted signal. To do this, the receiver makes use of multiple filters of various different technologies. A perfect filter would allow through, without attenuation, the wanted signal, but attenuate all other signals to such a large degree that they would have no impact. Such a filter is often described as a "brick wall filter" because when its characteristics are plotted in the frequency domain it is rectangular – it looks like a building with vertical brick walls.

Such filters cannot be practically implemented. Instead they are all somewhat imperfect, attenuating the wanted signal slightly and allowing through some unwanted signals, especially in the immediately adjacent frequency bands. Better filters can be achieved with more components, or in extreme cases using low-temperature superconductors, but these approaches are expensive and might add unnecessary cost to consumer items. Hence manufacturers typically make receivers that they consider are just good enough to work in practice.

A receiver will have a figure of merit known as its adjacent channel selectivity (ACS). This is the level of signal on an adjacent channel that can be tolerated before interference occurs. A typical level might be, say, 30 dB. This means that the adjacent channel signal can be 30 dB greater (1,000 times greater) than the wanted signal but that after filtering in the receiver its strength is reduced to a level that is no longer harmful. However, if the MCL is small then this can still be problematic. Imagine a TV receiver that can decode a wanted signal which is as weak as −80 dBm. This means it could tolerate a signal in the neighboring band at about −50 dBm (assuming a 30 dB ACS). However, if the neighboring band was used by a cellphone with a 100 mW transmitter (20 dBm) and the MCL was, say, 40 dB for both in the same building, then the signal from the phone would be at −20 dBm at the TV receiver – much larger than could be tolerated.

Spectrum managers would prefer receivers to have very high ACS figures. Manufacturers would prefer that spectrum managers do not give out licenses that might result in high levels of interference on neighboring bands. At present, the manufacturers tend to win out as there is no easy way for regulators to enforce receiver standards. However, this is a topic of much discussion where there may be change in the future. We discuss this further in section 10.3.

The interference experienced is a combination of the two and in some cases one can tend to dominate. For example, if the ACL of the transmitter is 50 dB below the wanted signal, then with a 1 W transmitter and 40 dB MCL the interference in the band will be −60 dBm. If the ACS of the receiver is also 50 dB then the same −60 dBm signal will be received, adding to a total interference of −57 dBm. But if, say, the ACS of the receiver was only 40 dB then this would dominate the interference and the transmitter ACL would be irrelevant. In a perfect environment, a regulator would aim for these two mechanisms to be approximately equal.

2.4.5 Blocking and receiver issues

Other problems can occur in poorly designed receivers. A particular concern is "blocking." This is when there is a very powerful interfering signal some distance away in frequency from the wanted signal. Normally the filter in the receiver would remove this signal. But it may be so powerful that it overloads the components prior to the filter (often a low-noise amplifier), resulting in their performance degrading. This can be a particular problem with TV reception where there are amplifiers near the antenna (e.g. in the attic of a building). These external amplifiers typically have very limited filtering as they aim to amplify all the different TV signals available, allowing the various TVs in the house to then select the particular one of interest to them. A powerful interferer somewhere in the TV band would be amplified by this device and may result in a signal so powerful it overloads the amplifier. There is little that can be done about this other than taking care in designing receivers.

Some receivers can also be particularly sensitive to interference on certain channels. For example, many receivers change the frequency of the wanted signal to a fixed "intermediate" frequency. This can be done using a mixer and is advantageous in that very tight precise filtering can be implemented at the intermediate frequency. Mixers actually result in two different signals being translated to the intermediate frequency at equal offsets above and below the mixer frequency. One will be the wanted signal. If there happened to be interference at the other offset this would be translated directly into the intermediate frequency and be particularly troublesome. Many TVs have receivers which use a similar approach and as a result are vulnerable to an "$n + 9$" interference – meaning a signal on a TV channel separated by nine channels (72 MHz in Europe) from the wanted signal.

2.4.6 Unusual cases: AGC and similar

Finally, unexpected interference can occur due to particular design choices. A recent case has been interference between non-TV transmissions (both cellular and "whitespace" devices such as machine communications systems) and TV receivers.

In most receivers there is a circuit that adjusts the level of amplification of the incoming signal such that the amplification is reduced when the signal is more powerful. This ensures a near-constant signal level at the decoding circuitry. This circuit is called an automatic gain control (AGC). One design parameter is the speed at which it reacts.

Too slow and signal levels may go out of tolerance, too fast and it may try to respond to short fluctuations that are not important. Until recently, the TV band was used almost exclusively for TV signals. These are near-constant in power and so the environment changed only slowly. A slow-changing AGC was suited to this. But systems like LTE generate rapidly changing bursts of transmission. This "bursty" signal caused the AGC to start to react slowly and to continue to react after the transmission had stopped before finally coming back to its starting point just about the time the next burst occurred. The forever-changing gain tended to confuse the TV receiver. So while the receiver could tolerate a constant interferer on a neighboring channel it could not tolerate a bursty one. TVs are now being designed to cope, but there are many legacy units that will take a decade or more to be replaced.

Other unexpected interference cases are likely to occur as a wider range of different technologies and usages share spectrum.

2.5 Tolerance of interference

2.5.1 Overview

Interference occurs in all radio systems – even without man-made transmissions there is interference from thermal noise and solar-radio effects. Then there are unwanted emissions from electrical devices and finally other interfering radio emissions. So all systems need to tolerate some level of interference – and the more they can tolerate the more flexible their deployment and the higher the efficiency of spectrum usage that can be achieved.

There are many ways that system designers can reduce the susceptibility of their devices to interference. Many of these were mentioned above in describing the different sources of interference and are developed further in this section.

2.5.2 Improved receivers

Improving receivers can help hugely. In particular, interference is often dominated by the receiver ACS. It has been this that has led to major issues in the US surrounding deployments like LightSquared's proposed broadband system adjacent to the GPS band (see section 10.3.2). Studies for Ofcom [6] have shown that for an additional cost of about a dollar most receivers can improve their ACS by around 6 dB. This can make a huge difference to the viability of the use of spectrum by other systems. Yet many receiver manufacturers are seeking further economies that worsen the performance of the receiver.

2.5.3 Frequency hopping and other dynamic approaches

Another option is to work around any interference. Interference can often only be on a few nearby channels and can change quickly as devices move around. For example, the interference to a TV from a cellphone might change from second to second as the phone moves past the house, periodically retrieves data, then becomes quiescent, and so on. If a device has multiple channels available to it, then it can continuously hop from one to another. If there is interference on one, the effect will be short-lived and typically can be overcome using error-correction systems. This is the approach adopted by Bluetooth devices to overcome interference in the unlicensed 2.4 GHz frequency band that they use.

Systems can also adapt their behavior – for example, reducing data rates while there is interference such that more interference can be tolerated. More advanced systems make use of multiple antenna systems (often called "MIMO") which can steer their directivity away from interferers, creating a "null" in the direction of the interferer and so much reducing its effect.

2.5.4 Difficulty in planning: at800 experience

With all these variables it is unsurprising that planning a licensing system to optimize interference and capacity is nearly impossible. Quite how difficult this is was brought to light recently in the UK.

LTE, or 4G cellular, is being deployed in a band directly adjacent to TV transmission. Modeling work had suggested that this would cause interference to TV reception because the TV receivers would be poor at filtering out this adjacent band signal [7]. After much consultation and research, the UK regulator, Ofcom, concluded that around 2.3 million homes, about 10% of the population, would be affected. As a result, a £180 million campaign called "at800" was put in place to alert viewers, provide them with filters, and, if that failed, to visit their homes and assess the alternatives open to them.

It was decided to run some pilot deployments to check that the process was working correctly. In these pilots, hardly any viewers complained of issues [8]. As a result, Ofcom has now revised down the likely number of affected homes from 2.3 million to around 50,000 – only some 2% of the initial estimate. Even this may be too high. Clearly the modeling work was hugely pessimistic and, if applied generally, would result in much less spectrum being made available than could be safely used.

How could this be the case? The analysis of where the modeling went wrong is still under way but it appears to be the general problem of adding conservative assumption to conservative assumption. Each assumption is reasonable but in aggregate the chances of them occurring is very limited. There may also be more subtle factors at play. For example, the areas where interference is most likely to be problematic are where TV signal levels are low. But it is now thought that in these areas viewers may have already moved to other platforms such as cable or satellite because of the poor reliability of the terrestrial TV signal, in which case they would not notice any interference.

This strongly suggests that models are unlikely to capture all the subtleties of the real world and that a better approach may be one where extensive trials and ongoing adjustments to transmitter levels are made as experience is gained.

2.6 The need for regulation

2.6.1 Where interference cannot be avoided

Throughout this chapter we have referred to the role of the regulator. If there were no interference and all the users who wished to transmit could be accommodated then there would be little need for regulation. Everyone could just do whatever they wanted free from bureaucracy. The general view is that we are far from this situation and that, if left unregulated, there would be massive interference between competing users that would reduce the value of the spectrum. In such a world the task for the regulator is to manage the interference in such a manner that the maximum economic value is gained from the radio spectrum.

Some would debate this and argue that in unlicensed bands, such as at 2.4 GHz, the regulator does not control access to the spectrum and yet devices work well, without interference, generating substantial economic value and innovation. We will return to "commons" bands in Chapter 8, but broadly the regulator sets down the rules of access to the band but then does not license individual users. The rules of access always include power limits that are typically set low so that interference does not travel far. (This is the reason why the range of Wi-Fi devices is much less than that of cellphones.) The combination of high frequencies, low transmit power, and use that is often indoors and so provides some shielding from other users means that the interference potential is low. When this is coupled with design approaches like frequency hopping, the result is that broadly the systems work – although increasingly in some cases congestion of Wi-Fi networks can be problematic.

However, this laissez-faire approach breaks down as the range of systems increases and so their potential to interfere grows. It is also problematic for those deploying large and expensive networks that require long-term certainty of interference-free access in order to justify the up-front expenditure. This is why networks such as broadcast TV and cellular are licensed and likely to remain so for the foreseeable future.

So the role of the regulator is, then, to issue to these operators licenses which contain parameters such as allowed transmit power and transmitter masks in such a way as to minimize interference while not resulting in overly complex equipment or unnecessarily large guard bands between users.

2.6.2 Difficulties of optimal regulation

Summarizing this chapter, transmitters generate radio signals with particular parameters, including power, bandwidth, and adjacent channel leakage. These signals propagate through a complex environment to receivers – both those that want to receive this signal and those that are trying to receive a different signal. For those receiving a different signal the transmission will appear as interference through a number of causes, including the in-channel ACL and the neighboring-channel emissions that the receiver does not filter correctly. Whether this interference will be harmful in any way depends on many parameters, some of which are not under the control of the regulator.

All of this makes it clear that it is just not possible to come up with perfect "optimal" regulation that ensures that all interference is just below the threshold of being noticeable while allowing the entire spectrum to be fully used. If the regulator were to aim for this perfect point, then in some cases they would overestimate interference, resulting in conservative usage of spectrum, and in some cases they would underestimate, resulting in deployed systems experiencing interference that might be harmful. Most regulators judge the latter to be more problematic than the former – if nothing else because it will result in more complaints and investigations of their behavior. Hence they tend to aim for a point some way below optimal regulation using conservative licensing. Recent experience suggests that by adding up multiple conservative assumptions, this point is actually a long way below optimal.

How this might be addressed is the topic of much of the rest of this book.

References

[1] C. Elwood Shannon, "A Mathematical Theory of Communication" (July–October 1948) 27 *Bell Systems Technical Journal* 379.
[2] C. Haslett, *Essentials of Radio Wave Propagation*, Cambridge University Press, 2007.
[3] Ofcom, "Spectrum Framework Review" (June 2005), at http://stakeholders.ofcom.org.uk/binaries/consultations/sfr/statement/sfr_statement.
[4] See http://en.wikipedia.org/wiki/Hata_model_for_urban_areas or any text on propagation.
[5] See www.radartutorial.eu/07.waves/wa17.en.html for more detailed explanation.
[6] Ofcom, "Study of Current and Future Receiver Performance"; see http://stakeholders.ofcom.org.uk/market-data-research/other/technology-research/research/spectrum-liberalisation/receiver.
[7] See http://stakeholders.ofcom.org.uk/binaries/consultations/dtt/annexes/lte-800-mhz.pdf.
[8] See www.fiercewireless.com/europe/story/uk-industry-group-reports-minimal-lte-interference-digital-tv/2013-04-05.

3 The economic challenge
A basic primer on spectrum economics

The radio spectrum is a resource of great significance to all modern economies. The importance of services supported by radio spectrum has grown markedly in recent years, especially as more and more mobile communications applications take hold among the world's population. It is thus critical that this increasingly important resource is allocated efficiently, in a way which maximizes the benefits which people gain from their individual use of services such as mobile telephony or broadcasting, or which they gain collectively from the availability of spectrum using public goods such as defense.

The aim of this chapter is to set out the requirements for efficient use of spectrum and, in particular, to show how methods of spectrum management based on the use of markets or price-type instruments can help in this goal.

3.1 Characteristics of spectrum as an economic resource

Spectrum can be viewed as if it were a finite resource, in the sense that in a given frequency band within a specific location there will be a physical limit to the amount of use possible. In this regard a frequency band is analogous to land – its location and area is given but its productive capability is dependent upon technology. Radio spectrum is heterogeneous, as different frequency bands are suited to different purposes, as is illustrated in Table 3-1, although there is scope to some degree for substitution of one band for another in certain uses.

Unlike oil or gas, it is not storable: if a band is not utilized over any period the opportunity to use it can never be recovered. Equally, it is not exhaustible: it cannot be depleted or used up. It has to be used in its existing location or not at all; there is no possibility of importing or exporting it.

More can be got out of spectrum in two ways: first, by bringing into use bands which have not been used before; in the case of land, where this is also true, this has been known as the "extensive margin." Equally, as also with land to which fertilizer is applied, spectrum can be made more efficient by such things as compression techniques, which increase efficiency on what can be called the "intensive margin."

By way of comparison, Table 3-2 shows how spectrum compares in these and other respects with land, oil, and water. A further key feature of spectrum is that, as noted in Chapter 1, because of interference, one use of spectrum (say, mobile telephony) can interfere with another use (say, broadcasting) in a very complicated way.

Table 3-1. Bands and their usage

Band	Frequencies	Services (a few examples)	Approximate time use began
Medium frequencies	300 kHz–3 MHz	AM radio	1920s
High frequencies	3 MHz–30 MHz	Shortwave radio	1930s
VHF (very high frequency)	30 MHz–300 MHz	FM radio, broadcast TV	1940s
UHF (ultra high frequency)	300 MHz–3 GHz	Broadcast TV, mobile and cordless phones, Wi-Fi, Wi-Max, paging, satellite radio	1950s
SHF (super high frequency)	3 GHz–30 GHz	Fixed microwave links, Wi-Fi, cordless phones, satellite TV, wireless fibre	1950s 1970s
EHF (extreme high frequency)	30 GHz–300 GHz	Short-range wireless data links, remote sensing	1990s

Table 3-2. Characteristics of different natural resources

	Spectrum	Land	Oil	Water
Is the resource varied?	Yes	Yes	Not very	Not very
Is it scarce?	Yes	Yes	Yes	Yes
Can it be made more productive?	Yes	Yes	Yes	No
Is it renewable?	Yes	Partially	No	Yes
Can it be stored for later use?	No	No	Yes	Yes
Can it be exported?	No	No	Yes	Yes
Can it be traded?	Yes	Yes	Yes	Yes

Source: [1].

Some of these features simplify the spectrum management process considerably. For example, the fact that spectrum is non-storable means that we do not have to worry about whether to save it to use later. Thus there is no problem, as there is with oil, for example, whether we should use it now or save it for more valuable later use. On the other hand, the interference to which adjacent spectrum users are subject makes it vital to protect one user from others.[1] If this condition is not satisfied, spectrum might be rendered entirely useless by interference.

---

[1] Similar issues arise with land management, where use of land, for example for a chemical plant, may interfere with use of adjoining land for another purpose, for example a residential estate.

3.2 What is an efficient allocation of spectrum across uses?

The real problem of spectrum allocation is extremely complicated: how to allocate a large number of different bands to many different known and unknown future uses, when the demand for such uses is growing at variable and unpredictable rates and when new uses are constantly coming into existence.

We begin the discussion of an efficient allocation of spectrum, however, using a highly simplified example: how to allocate a given amount of spectrum in a single band across two uses, when no other band will do and where demand for the two services which use the spectrum is known and invariant.

This situation is illustrated in Figure 3-1. The total spectrum available for allocation in a band is shown on the horizontal axis. Any point on the line shows an allocation between two uses, service 1 and service 2. Thus the left-hand side shows 0% used in service 1 and 100% in service 2; the right-hand side shows 100% used in service 1 and 0% in service 2. For example, the two services can be mobile communications and television broadcasting.

The curve originating on the left-hand vertical axis shows the marginal benefit of spectrum for service 1. The idea behind this is that a small amount of spectrum allocated to service 1 will permit an initial tranche of (for example) mobile calls to be made, and these calls will yield considerable monetary benefits for consumers – which are demonstrated by their high willingness to pay for such calls. (Thus they might be emergency calls or important business calls.) As more spectrum is allocated, the additional or marginal benefit of such calls declines, eventually getting down to zero.[2]

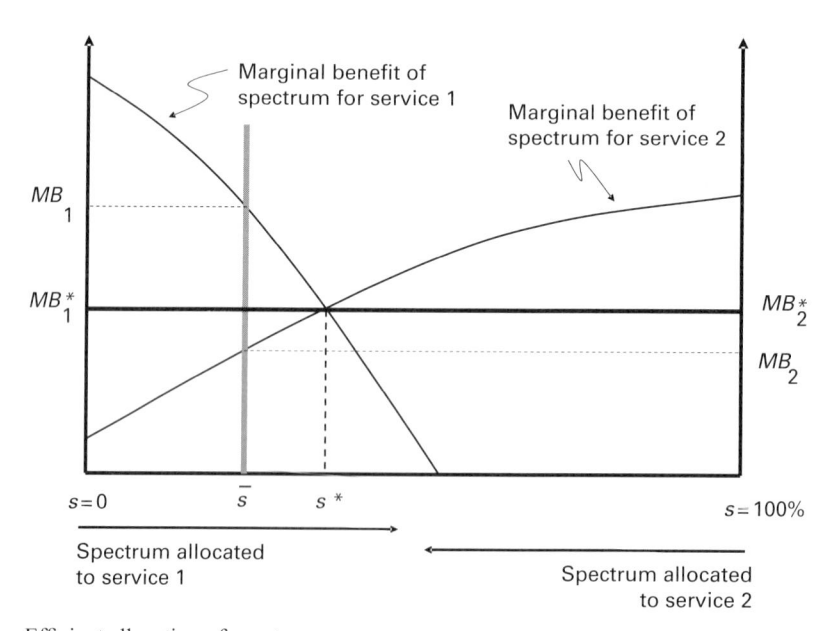

Figure 3-1. Efficient allocation of spectrum.

---

[2] Note that here we are ignoring the "network effects" of mobile communications.

Equally, the curve originating from the right-hand vertical axis shows the marginal benefit from spectrum for service 2. If service 2 is, for example, pay-TV broadcasting, the declining shape of the curve reflects the fact that viewer benefits from an initial TV channel are high, but the additional or marginal benefit of further channels decreases as they are successively added. (Note that spectrum allocated to service 2 is measured from right to left.)

How should spectrum be allocated between the two services? As is well understood in economics, an efficient resource allocation occurs when marginal benefits are equal. Given this, an efficient allocation occurs at point s^* on the horizontal axis where the two marginal benefit curves cross or the marginal benefit of spectrum in each use is the same. At this point, taking a unit of spectrum for service 1 and reallocating it to service 2 has no effect on total benefit. Whereas at point $\bar{s}$, for example, the marginal benefit in service 1 exceeds that in service 2, so a reallocation in favor of service 1 will increase total benefit from the band.

This reasoning enables us to identify the condition which an efficient allocation of spectrum will satisfy. We now go on to consider, first, more realistic complications, and then how alternative spectrum management regimes can seek to get close to the optimum.

3.3   A more realistic formulation of the problem

So far we have considered the simplest case: one spectrum band with no substitute, two uses, and known invariant marginal benefit curves. In practice, the problem is much more complicated, in the following ways:

- There are many frequency bands. Their usage reflects existing regulatory arrangements, made globally, regionally, and nationally. This multiplicity makes the efficient allocation problem much more complex.
- There are many potential uses within a single band. Thus high-value bands, lying between 300 MHz and 3 GHz, can be used for a variety of mobile communications uses, for television broadcasting, for radar, and for a number of other uses.
- Substitution between bands is possible. For example, mobile communications can be and are provided using a range of frequencies from 300 MHz to 3 GHz and up. But they are not perfect substitutes. Instead the frequencies differ in propagation characteristics, building penetration, etc. Taking into account substitution possibilities across bands means that the problem of allocation in each band has to take account of what is going on in substitute bands as well.
- The benefit curves are not generally known. Essentially they depend on two things – the willingness of end users to pay for different quantities of the services associated with the spectrum uses, and the degree to which the spectrum in question can be substituted for by other inputs in production of the services.[3] Some of this knowledge may be held by the firms producing and marketing the services, but it is not common knowledge.

---

[3] The first arises because spectrum, as an input, is not wanted for its own sake but demand for it is derived from the services it can produce. The second consideration means that the price at which a substitute is available acts as a cap on the value of spectrum in any use.

- The benefit curves are shifting all the time. Whenever either demand for a service or the costs of producing it change, the benefit curve shifts. Even more disruptive is the arrival of completely new services – in the ICT world this can occur rapidly and the services become very widely used in a relatively short period. The new services may also replace existing ones. This creates a severe challenge, because spectrum allocation has traditionally been a fairly deliberate or even a slow process, it being necessary to clear bands used for one purpose before they can be used for another. Thus an efficient allocation of spectrum requires the exercise of some foresight – or alternatively should allow for sufficient flexibility to cope with unforeseen events.

Solving the allocation problem moderately efficiently in a centralized fashion was conceivable in the past, when spectrum had relatively few uses and demand could be accommodated easily within available frequency bands. But performing the task in today's world would require the collection and processing of an enormous amount of information. It is therefore natural that consideration has been increasingly given to decentralized or delegated methods of allocation. We consider these below, after a fuller discussion of some of the many possible interpretations of "access to spectrum."

3.4 The broad range of modes of access to spectrum

In recent years, it has been recognized that, before spectrum is assigned to users, decisions have to be made over what is meant by "access to spectrum." The traditional answer is that "access to spectrum" means that an operator has for a predetermined period of time an exclusive right of access to a specified band in a particular geographical area. However, the range of modes of access to spectrum has always been broader than this, and that range is now expanding rapidly with the emergence of new methods of sharing access to spectrum.

The most prominent alternatives in the designation of rights of access to spectrum are set out in Figure 3-2. At the very top of the figure stands a public commons, a form of

Public commons Private commons Dynamic spectrum access	**Rules-based**
Vertical sharing (e.g. white spaces or underlays)	**License- and rules-based**
Licensed/authorised shared access Exclusive shared access	**License-based**

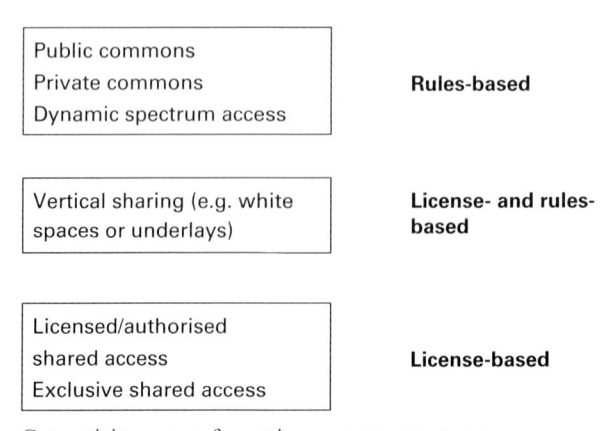

Figure 3-2. Categorizing ways of ensuring access to spectrum.

open access for all users satisfying certain regulatory rules. Here are also found the private commons, open to all users admitted by a firm or organization, and dynamic spectrum access, a new mode which is able to take account of congestion. At the very bottom we find exclusive licensing, which, for a period, which may be infinitely long, gives a user uncontested access to a given band in a specified geographical area. In this area we also find access shared between two or more identified parties, with clear rules of division or a clear hierarchy of rights of access. In the middle we find a combination of privileged users and unknown others with secondary rights of access.

The non-exhaustive list of sharing options between the extremes of a public or private commons and an exclusive license can be summarized as follows:

- In a private commons, where an exclusive license is made available to a private operator, which then gives access to it to specified users (for example, all those who buy a particular device), all qualified users have rights of access, without a real-time means for preventing congestion. Lack of interference thus depends on the stipulation of rules of use, such as the power level of transmissions.
- In dynamic spectrum access, all qualified users can seek access to spectrum, but first have to inquire which particular frequency is free from interference; in other words, there is a real-time regime for avoiding interference.
- With vertical sharing, a licensee is identified which typically has a prior right of access, but other users obeying certain rules requiring them not to interfere with that licensee or with one another can also be accommodated.
- In the case of shared access, two or more licensees are identified, which share access to the spectrum on a geographical or temporal basis, or, for example, by giving one user priority over another.

3.5 Alternative ways of allocating and assigning spectrum

Throughout history, different ways have been found of allocating economic resources. At one extreme, all decisions can be centralized and implemented by issuing orders to producers as to what must be produced and with what inputs. This can be combined with directly allocating consumption goods and services to households. At the other end of the range of possibilities, all resource allocation can be done through markets, with no central direction.

At the level of the economy as a whole, most countries do not choose points at these extremes. Thus even in market economies, governments make many resource-allocation decisions by collecting taxes and making spending decisions, by producing public services and by retaining regulatory powers to direct or influence how resources are used. Equally, it has proven impossible in the practical operation of a modern economy for the government literally to control all allocation decisions. Most have to be delegated to some degree to producers and consumers.

The same mixed regime applies to many natural resources. Thus oil and gas are regulated in terms of their extraction and production, often taxed, and sometimes owned by public firms, but then normally allocated by a market process. Land is frequently restricted in terms of its permissible uses by zoning laws, but is generally traded in various forms, including freehold and leasehold. A natural resource which is very often allocated by command and control is raw water extracted from rivers or aquifers. In many countries abstraction rights are allocated by command and control to firms responsible for the public water supply or for such things as electricity generation. In other countries, especially where water is in short supply, a market for water operates.

As far as spectrum is concerned, allocation has relied very heavily – far more heavily than allocation of land, for example – on direct command-and-control allocation. The rationale for such intervention, orchestrated at a global level through the ITU, was to minimize interference, which can be very costly, and to manage risks by providing greater assurances to firms manufacturing radio equipment. Consequently, at a very early stage and prior to the explosive growth of radio-based personal communications, the ITU recommended the allocation of many frequency bands to specific uses – allocating swathes of frequency in the 300 MHz–3 GHz bands to broadcasting and defense-related activities. Most national governments have adhered to the ITU recommendations and, furthermore, assigned licenses to users within these bands, often accompanied by a detailed specification of the services to be provided, what equipment is permissible, at what power levels, and so on.

In terms of our previous account of the spectrum allocation problem, an efficient allocation of spectrum by command and control requires the spectrum regulator to know, or to act as if it knew, the present and future benefits of its uses in different areas. This is obviously a tall order. It is thus pertinent to ask why command and control has predominated in this way.

The possible reasons for the high degree of centralization of spectrum management, compared with other inputs, include:

- The highly technical nature of the input and the technologies used to exploit it.
- The link between spectrum and national security; at the time of World War I, this led to the assumption of public control over spectrum management which has persisted to this day.
- The complexity of preventing interference among spectrum users; a regime of command and control which specifies the band to be used, and the location and power of transmitters, is one way of controlling interference.
- Before the large expansion of demand for wireless communications, the lack of shortage of supply of spectrum; this is probably key, as a good supply of spectrum relative to demand reduces the costs of any mistakes or tardiness in allocating spectrum.

However, recent developments have challenged the old ways. The main factor changing the situation is the increasing pressure placed on demand for spectrum, first by the massive growth of mobile voice usage, now by equally large recent and projected growth rates of mobile data services. These developments have made it more important than ever to use spectrum efficiently and also to adapt quickly to the new pattern of demand. A further major factor is the fast pace of technical change which already permits or will permit new ways of using spectrum based on a much greater degree of sharing.

What is the principal alternative to command and control?[4] In very general terms, the conventional answer has been the use of spectrum markets. Essentially, this turns over the allocation problem to firms and, indirectly, end users of spectrum-based services. It is accomplished by creating, or allowing to come into existence, a marketplace in which spectrum is available for sale and purchase.

Under this approach firms formulate their strategies concerning spectrum use on the basis of their costs of production of the spectrum-using service which they produce and their estimate of end users' demand for it. All firms then go into the market either to buy or to sell spectrum in a particular band or bands at the prevailing market price. If a willing buyer can find a willing seller in this market, a market transaction can be accomplished. The process is intended to work by delegating decisions to a group of firms, each of which brings to bear its own private knowledge. As discussed in Chapter 12 below, either public-sector spectrum can be allocated separately by traditional command-and-control methods, or both public- and private-sector spectrum markets participants can trade with each other in an integrated market.

More generally, it is unlikely that a spectrum manager will want to adopt a "one-size-fits-all" approach, applying a uniform instrument of control to uses of spectrum which include military defense, international satellites, national mobile communication operators, and private mobile radio used by local taxi firms. It is more likely that different bands will require different approaches.

As the following Box 3-1 explains, the idea of creating a market for spectrum has been around for over fifty years. It was first put forward in detail by Ronald Coase in a paper published in 1959.

As a result of Coase's work, spectrum became an important example of the alleged benefits of the development of property rights in assets in general. Thus Demsetz argued that, as against the alternative of allowing open access to an asset by all parties, the application of private property rights was more efficient, because private ownership gave an incentive to protect and improve the asset [2]. In addition it was pointed out that efficiency could be enhanced by trading – allowing the asset to be transferred to a new owner who can use it more efficiently. The comparative assessment of ownership models was extended to include exclusive ownership, shared ownership, public ownership, and the periodic auctioning of access rights.

Rather than pursue the general argument about the merits or otherwise of private property rights, we will focus on the specifics of arrangements for access to spectrum, which, as the previous section has shown, create an unusually large range of technical options, many of which, such as "white spaces" and other forms of shared access, have only emerged in recent years.

---

[4] We have already noted the third option of unlicensed spectrum or the commons in section 1.1.3 above, and will return to this in more detail in Chapter 8 below.

> **Box 3-1.** Coase's 1959 idea for a spectrum market and its reception
>
> In 1959 Ronald Coase, an English-born economist at Chicago University, awarded the Nobel Prize in economics in 1991, published an article called "The Federal Communications Commission." In the discursive style then adopted by economists interested in legal issues, it raised the issue of replacing the then-current practice of the US communications regulator of assigning TV licenses by expert selection with use of a market system.
>
> The reaction to the work was very hostile. Some readers treated it as a joke. Some foresaw the end of civilization. Several argued against its publication. For reasons which now seem obscure, it elicited an almost universally violent reaction.
>
> The idea of access to spectrum as an asset to be traded was not taken up for another 25 years, when wireless license or spectrum auctions began, first in New Zealand, then in the USA, then in Europe and everywhere else. This was followed by the use of other market methods. As is shown below, the various strands of use of market- and price-based allocation methods for spectrum allocation are used unevenly, but the idea is everywhere on the policy agenda.
>
> Sources: R. Coase, "The Federal Communications Commission" (1959) 2 *Journal of Law and Economics* 1, available at www.eecs.berkeley.edu/~dtse/coase.pdf. R. Coase, "Comment on T W Hazlett, 'Assigning Property Rights to Radio Spectrum Users. Why Did FCC Spectrum Auctions Take 67 Years?'" (1998) 41 *Journal of Law and Economics* 577.

We return to these modes of access in more detail in Chapters 8 and 9. But before doing so we briefly review the four forms or dimensions of market or price-based transaction considered below. These are:

- Auctions: mechanisms used to make an initial assignment of spectrum licenses. Instead of choosing the identity of licensees of mobile or other spectrum through a comparative selection or beauty parade, the spectrum regulator could auction licenses, subject to requirements setting out what service had to be produced and how. This approach has the additional attraction of raising revenue for government and ought to ensure that license(s) are assigned to the most efficient operator(s).
- Liberalization, or giving the licensee more choice in the technology it uses or the service it produces. In this option the licensee is given the choice over what technology it uses (e.g. 2G, 3G, 4G) and possibly over what services it produces. The latter freedom creates major issues over policing interference, discussed further in Chapter 10.
- Spectrum trading, or introducing a full secondary (resale) market for spectrum licenses. In this option, as well as being auctioned at the start of the license, the spectrum (or parts of it) can be bought and sold or leased in the course of the life of the license.
- Administrative pricing: the setting of a (usually non-zero) administrative price to accompany direct spectrum assignments.

3.5.1 Auctions

Auctions have taken over as the major means of assigning mobile telecommunication licenses throughout the world. Over the last 25 years or so, they have effectively replaced the previous method, known as "beauty contests" or comparative administrative procedures, in which a body is formed which assesses rival applications, normally using criteria which are published in advance.[5] The license or licenses are then assigned without a monetary payment.

Auctions are discussed and evaluated fully in Chapters 4 and 5 below. Here we make a number of more general points bearing on the comparison of auctions and beauty contests for new spectrum awards.

The first is that a beauty contest without a monetary payment risks over-rewarding the successful applicants. To see this, consider how a firm decides how much it is prepared *at a maximum* to bid for a mobile license. The firm will calculate the excess of its expected revenues over its expected costs, excluding the cost of acquiring the license, for each year of the term of the license. The net present value of this stream is calculated by discounting the annual payments at an appropriate rate of interest. That sum should be the maximum a firm would be prepared to bid. It will hope to get away with less, but the more competitive the auction is, the closer it will have to go to that amount to get the license.

In a beauty contest, by contrast, the firm which is chosen will be able to keep the whole of that excess. In other words, an auction will transfer to the government the bulk of the so-called rent or excess profit associated with the license, while a beauty contest will leave the rent associated with that contest with the firm.

This aspect of auctions has its risks, however. If more spectrum is available to provide services, such as mobile communications, and if the most efficient use of that spectrum is in the sector, then it is desirable that the spectrum be put to use there. However, it might occur to the government that rationing the supply of spectrum would enhance revenues from spectrum auctions. That is, a government might create a deliberate shortage to put up the auction price and enhance revenues.

This is extremely short-sighted, since a higher price of spectrum is based on firms' expectations of higher prices to end users. So rationing spectrum restricts take-up of mobile voice and data services, which in turn damages a country's economic prospects. In other words, it is desirable to extract rents from bidders in auctions where spectrum supply is not artificially distorted, but undesirable to take this too far if it involves artificially restricting spectrum availability.

Second, although beauty contests are intended to assess the likely conduct of each applicant if it were granted the license, inevitably the data provided are in the form of future commitments rather than proof of current or past actions. As a result, applicants are inclined to overpromise, hoping to escape at a later stage from commitments made at the time of the contest. As a result, there is a risk of problems later. For example, Norway

---

[5] Beauty contests are still sometimes employed, and mobile operators, for obvious reasons, frequently support them.

allocated four 3G licenses using a beauty-contest approach. One of the winners (Enitel) became insolvent and another (Tele2) returned its license after being unable to meet the network deployment commitments that it had given.

In an auction, by contrast, it is possible to extract the financial sum promised by the successful bidder by requiring it to be paid before the license is granted. In these circumstances, it would be reasonable to hope that, if the auction is competitive, the licenses would go to the highest bidders, which would also be the most efficient operators. This would be a very satisfactory outcome in terms of creating an efficient sector and in gaining revenue for the government.

Unfortunately it will not always work out this way. This arises because of a phenomenon known as the winner's curse. Thus a firm might gain a license because it is one of the most efficient applicants for one (a good reason), or because it has the most optimistic expectations (a bad reason, if the expectations are unrealistic). However, auctions can be designed in ways which seek to deal with problems of the winner's curse, as will be explained in Chapters 4 and 5 below.

Third, as compared with a beauty contest, an auction involves less discretion in the assignment process, so is less vulnerable to corruption.

Fourth, an auction for spectrum licenses can have harmful consequences of an anticompetitive kind. A dominant mobile firm or a combination of large firms can use their deep pockets deliberately to deprive smaller competitors of access to spectrum by acquiring more than they need. (This can be, and has been, countered by the imposition of spectrum caps, which limit the amount of spectrum which a single firm can buy at an auction, or can hold in aggregate.) Or a larger group of rivals can agree not to bid against one another, and thereby collusively keep prices down.

Despite these risks, and the additional risk of the auction going wrong because of technically poor design, use of auctions is now the default approach in making awards of new spectrum. A summary of their pros and cons is given in Box 3-2.

3.5.2 Liberalization of spectrum use

By liberalization, we mean allowing the spectrum licensee more discretion over how it uses the spectrum which it has been awarded. The two main elements are:

- "technological neutrality," or allowing the licensee the choice of what technology to use to provide a specified service (e.g. 2G, 3G, or 4G mobile), and
- "service neutrality," or allowing the licensee to decide what service to provide (e.g. digital terrestrial television or 4G mobile).

Thus technological neutrality allows a mobile operator to vary the technology it employs over its portfolio of bands as customer preferences change, without having to seek permission from the regulator. Because the service is unchanged, the interference characteristics are unlikely to vary very much.

It is very hard to see how restrictions on technological neutrality can benefit end users. In particular, they can impede fundamentally straightforward technical developments, such as transfer of 2G spectrum to 4G spectrum.

> **Box 3-2.** The pros and cons of auctions
>
> **Pros**
> - A properly devised auction should direct the license to the most efficient operators, as they can afford to bid higher.
> - A fully competitive auction will prevent the firms from making excess profits and redirect those profits to the government.
> - A well-designed auction is harder for interested operators or any other corrupt party to manipulate.
>
> **Cons**
> - In search of auction revenues, the government may artificially restrict the supply of spectrum, thereby damaging the country's economic prospects.
> - The largest firm or firms may buy up all the spectrum available to exclude competitors.
> - Bidders may collude to keep auction prices down.

The position is more complicated with service neutrality. If licenses are to be granted in a way which permits a change of service, a totally new interference management system may have to be introduced. Traditional interference management is based on assumptions as to what service a frequency is going to be used for, what technology (with what signal propagation properties) will be employed, where the transmitters are located, and what power they have. Using a computer model, the parameters of control of these variables are then carefully chosen to ensure there is no inappropriate interference with other users in contiguous geographies or frequencies.

But if the licensee can change at will the service delivered by the spectrum to one with different propagation characteristics and power, the system breaks down. In these circumstances an alternative has to be found. One such is the use of spectrum usage rights (SURs). As shown in section 10.2, this operates by imposing a limit on the interference which a spectrum user can project onto adjoining geographies or frequencies. Such new interference management regimes have been shown to work in some countries, but they are significantly more complex than the traditional ones. This is impeding the spread of liberalization through the introduction of service neutrality. Alternative means, such as administrative refarming, have to be used to redeploy spectrum.

3.5.3 Spectrum trading

In most countries, land can be traded at almost any time in many ways. Holdings can be broken up or amalgamated; land can be sold on a permanent basis or leased for a specified period; rights of use (e.g. farming and hunting) can be sold together or separately. The trading option for spectrum allows, in that natural resource, the same kind of flexibility as already exists with land in many places.

> **Box 3-3.** The pros and cons of liberalization
>
> 1. Technology neutrality
>
>   **Pros**
>   - Allows quicker adaptation of supply to customer demand.
>   - May reduce costs of supply.
>
>   **Cons**
>   - May create interference problems.
>   - Deprives regulator of lever (which may be misused) to influence operators.
>
> 2. Service neutrality
>
>   **Pros**
>   - Allows a wider choice of use of spectrum.
>   - Allows quicker transition in spectrum use.
>   - Enhances service competition.
>
>   **Cons**
>   - Probably requires major changes to interference regulation.
>   - Deprives regulator of lever (which may be misused) to influence operators.

Trades and other transactions in spectrum have to be recorded in some way in order to monitor interference. This can be done in a complex or a simple way. For example, in Guatemala, a pioneer of trading in the 1990s, a trade can be effected by a signature on the back of a license from the seller, which the buyer then takes to the spectrum regulator's office for recording.[6] In other jurisdictions, prior approval may be necessary.

The process of recording spectrum holdings can also disclose when a firm is acquiring a dominant position in any spectrum market. As noted above, there is a degree of substitutability across different bands for particular uses, such as mobile communications. But it is incomplete. For that reason, when spectrum caps are imposed for mobile communications, there may be separate limits on holdings of spectrum below 1 GHz, the most valuable spectrum, and for spectrum above 1 GHz. The caps can be absolute limits which cannot be exceeded in any circumstances, or "soft" caps, meaning that a proposed breach of them in an auction or by trading will trigger an inquiry.

Whether firms will exploit the opportunity to trade spectrum depends on the overall scarcity of spectrum and the interdependence of operators. A firm might be unwilling to sell to a competitor but be willing to sell unwanted spectrum to a firm outside its own market segment. Experience in trading is discussed further in Chapter 6.

---

[6] See Chapter 6.

> **Box 3-4.** The pros and cons of spectrum trading
>
> **Pros**
> - Allows spectrum to move more quickly between licensee and uses, without regulator involvement.
> - May enhance competition.
>
> **Cons**
> - Difficult to set up procedures.
> - May allow monopolization of spectrum.
> - Trades may not occur on a scale to justify set-up costs.

3.5.4 Administrative pricing

In command-and-control spectrum management regimes, spectrum is normally assigned to users by an administrative mechanism and a fee is levied, often based upon technical variables (such as power of transmitters, amount of frequency used, geographic area covered, etc.), aimed at recovering the administrative costs involved. This will usually mean that the charge for a license differs from the underlying economic value of the license and in most instances lies below it.

As we have seen above, the radical alternative to this approach is to create a primary market for access to spectrum (by use of auctions) and/or a secondary market by allowing trading, in which prices (the amount paid for accessing spectrum for a period of, say, one year) and valuations (the "capital value" of a license over its remaining lifetime) emerge from the interaction of supply and demand, as is the case, for example, with land in most jurisdictions.

However, an intermediate step between simple administrative assignment and a spectrum market is for the spectrum regulator to calculate and set prices which seek to reflect the current and near-term expected scarcity of spectrum. This provides the user with an incentive to economize on spectrum – an incentive otherwise absent if the price was zero or ignores scarcity.

This approach is usually intended to generate an annual charge for spectrum (a "spectrum price"). These prices can be aggregated to yield a capital value of access to spectrum over a run of years. We can thus distinguish prices which emerge from market operations from administrative prices which are set by the regulator.

Market prices reflect (but are not the same as) the willingness to pay for whatever attributes purchasers see in the product or service they are buying. In the case of spectrum, which is an input into production of other services rather than an object of desire in itself, willingness to pay for a spectrum license of, say 20 years, includes the following:

1. the degree to which holding the license will reduce the cost to the owner of producing the services the owner wants to produce,

2. the extra profits the owner will make as a result of the market power that ownership of the license will confer on it, and
3. an option value available to the operator at the end on the license if that operator is well placed to get a license renewal.

The sum of the factors above will normally yield a "maximum willingness to pay," but any buyer (for example a bidder in a spectrum auction) will want to pay less than this if it can. In other words, how the market works in practice, as well as underlying cost and demand factors, will influence the prices and valuations which emerge.

3.5.4.1 Opportunity-cost pricing

When a spectrum regulator sets administrative spectrum prices, it can perform the calculation to include any constituents that it chooses. For example, it can focus solely on the "cost-saving" element (1) above, asking, in effect, "if the mobile operator seeking to use this spectrum did not get access to it, and had to use the next-best alternative, by how much would its costs of producing the services it wants to produce go up?"

This measure is based on the so-called "opportunity cost" of the spectrum. Access to spectrum is priced on the basis of the extra costs which would be incurred if the spectrum were unavailable and, for example, the mobile operator had to use less spectrum and, in consequence, to install more base stations.

This magnitude also happens to be the price which would emerge for spectrum in a competitive market, so an administrative price set in this way mimics the operation of such a market. Suppose that, in a competitive market for mobile communications spectrum, one megahertz of 800 MHz spectrum is priced at $10 million a year more than one megahertz of 1800 MHz spectrum. On certain conditions this will reflect the fact that one megahertz in the 800 MHz band is associated with non-spectrum costs which are $10 million per year less than one megahertz in the 1800 MHz band.

Administrative pricing by the opportunity-cost method simply turns this relationship on its head. It sets the differential in administrative price per megahertz between the 800 MHz and 1800 MHz bands equal to the cost saving which the former offers compared with the latter. Sections 7.3–7.5 below set out the economic basis for this approach and give examples.

This approach should provide an incentive to firms to economize on spectrum use. It should also make it possible to capture for the government some of the scarcity value of unauctioned spectrum which an auction would have delivered had that market mechanism been employed. It also follows from this approach that if spectrum is in excess supply, its opportunity cost is zero.

There is growing experience of this method of calculating opportunity-cost prices, in a number of countries, including Australia and the UK. Initially, the estimation methods were fairly crude. For frequencies used for mobile communications, the alternative to more spectrum was taken to be the construction of additional base stations operating on lower power and permitting more frequent spectrum reuse. Now the focus is more on the use of alternative wireless bands. For higher frequencies, the construction of a fixed link was taken as the relevant alternative to microwave connections.

If analysis of this kind suggests that there is a misallocation of spectrum among uses, careful thought has to be given as to whether this is really a problem, and if so how it should best be solved. This is a necessary part of the decision process because, as well as setting administrative prices, the regulator could introduce trading or, if it knew the most efficient user of the spectrum, undertake a refarming exercise.

In its review of administrative pricing in 2009, the UK regulator Ofcom concluded that it was an effective instrument in certain conditions and that its use should be extended [3]. In 2011, the Australian regulator ACMA decided to use it widely [4]. But even in the UK, there is opposition from users.

The above discussion focusses on a particular way of calculating spectrum prices – via opportunity cost – normally for use in place of market prices or auction outturns to encourage economy in spectrum use.

3.5.4.2 The business-model approach to pricing

Opportunity-cost pricing is designed to capture what can be called the scarcity value of spectrum – the inherent difference across bands in their capacity to perform the service in question. For example, in mobile communications, lower-frequency spectrum can carry signals across a wider area (permitting economy in base stations) and can penetrate buildings better than higher bands. The scarcity of such bands makes them valuable, for the same reasons that fertile land is valuable. But access to such bands can also confer monopoly power (or, more generally, market power) in service markets, which adds further value to the spectrum. Some spectrum regulators will want to incorporate this element of value in their spectrum prices [5].

This aspect involves a different approach to pricing and valuation, which we call the "business-model" approach. It can be used, for example, to capture for the state some of the monopoly profits gained by firms through their access to scarce spectrum. It can also help in setting reserve prices in a spectrum auction – see section 3.5.1 above. Indeed, it mimics exactly the approach which, we suggested there, a firm would use in working out how much it was rational for it to bid in a spectrum auction.

Consider the firm bidding for a spectrum license at auction described in section 3.5 above. It may want to estimate its maximum possible bid by first working out the revenues it expects get and the costs (apart from spectrum costs) it expects to incur over the license; it then places a value on this stream of profits, taking account of the rate of interest it has to pay to borrow the money to finance the license acquisition. Of course, it will hope not to have to bid this entire amount, but to retain some of the profit for itself.

In setting administrative prices, a spectrum regulator could replicate this estimation process and set administrative spectrum prices which allow it to appropriate at least some of the excess profits based upon market power.

Making this calculation requires the regulator to project into the future not only the costs of providing the service (as is required with the "opportunity-cost" approach), but also the revenues. This requires assumptions about the competitive interactions among providers.

But while the task is difficult to perform precisely, it is not difficult to perform approximately, and to set prices which both recover some excess profit and avoid the risk (noted below) of pricing the spectrum so high that it falls out of use.

It is quite natural to use this approach when a spectrum regulator has decided not to re-auction spectrum at the end of a license period but to give the incumbent the option of "rolling it over" into a renewal without a competitive auction. (This might be done to avoid the risk of customer disruption following a change of operator.)

Obtaining a price or valuation of this kind may be of use in setting a reserve price in a spectrum auction, when there is doubt about the level of competition which might emerge – see section 5.4 below. In this case, the analyst might wish to prepare business models for a number of different firms. Then, if there are, say, four licenses, interest would focus on the willingness of the fifth-highest operator – whose bids the four successful firms would have to beat.

3.5.5 Summary

In summary, between traditional command-and-control methods of spectrum management and the use of auctions and markets, there lies the alternative of setting administratively determined prices for spectrum as a means of putting pressure on commercial and public-sector users to limit their demand for it.

Two methods of devising such prices have been discussed. The first prices a band by capturing its effect on the firm's costs. This method has been used in the UK in particular – with good effects, according to its regulator. The second method captures the benefits to the firm of any market power which a spectrum holding gives it; thus it may be used to help set a reserve price in a spectrum auction.

It is best to see administrative spectrum pricing as an additional tool used to encourage spectrum efficiency, probably in conjunction with other methods, in circumstances when auctions and markets cannot be used. Public-sector spectrum is likely to figure prominently in this set. Chapter 12 discusses in more detail the problems in principle and practice of pricing public-sector spectrum.

In order for the prices to have the desired effects, it is important to ensure that spectrum users themselves receive some reward for limiting their use of spectrum. This is necessary if they are to be willing to hand the spectrum back. It is also important not to set the prices too high, as this might have the effect of taking spectrum out of use altogether.

3.6 Conclusion

This chapter has shown that there are several ways in which market and price-based methods are being utilized in spectrum management. The most visible is the widespread use of auctions to assign licenses for mobile communications suppliers. This has taken hold throughout the world, and, as discussed in Chapters 4 and 5 below, increasingly complex auction formats are being used. Other regulators have gone further and

> **Box 3-5.** The pros and cons of administrative pricing
>
> **Pros**
>
> - Discourages spectrum hoarding and facilitates spectrum refarming.
> - Creates an incentive for more efficient spectrum use.
> - Captures some of the scarcity value of spectrum for the state.
>
> **Cons**
>
> - Method of calculating prices may be difficult to apply.
> - If prices are set too high, spectrum might be taken out of use.
> - Only works if licensee can benefit from reducing outlays on spectrum; this may not apply to public-sector spectrum users.
> - Faces strong resistance from those having to pay charges.

delegated more discretion to firms as to what technologies they choose to use and even what services they produce under their licenses. A further step is to allow the trading of licenses among firms in the course of their operation. There is also a role in some countries for administrative prices to be used to provide extra incentives in a command-and-control regime for users to economize on spectrum. Spectrum regulators should be aware of these approaches and consider them for application in their jurisdictions on the basis of specific uses and specific bands.

References

[1] ITU 2011, www.ictregulationtoolkit.org.
[2] H. Demsetz, "Why Regulate Utilities?" (1968) 11 (April) *Journal of Law and Economics* 55.
[3] Ofcom, "SRSP: The Revised Framework for Spectrum Pricing: Our Policy and Practice of Setting AIP Spectrum Fees" (December 2010), at http://stakeholders.ofcom.org.uk/binaries/consultations/srsp/statement/srsp-statement.pdf.
[4] See www.acma.gov.au/~/media/Spectrum%20Transformation%20and%20Government/Report/pdf/ACMA%20Response%20to%20Submissions%20Opportunity%20Cost%20Pricing%20of%20Spectrum%20Public%20Consultation%20on%20Administrative%20Pricing%20for%20Spectrum%20Based%20on%20Opportunity%20Cost.PDF.
[5] Aegis and Plum Consulting, "Estimating the Commercial Trading Value of Spectrum: A Report for Ofcom" (2009), at http://stakeholders.ofcom.org.uk/binaries/research/technology-research/specestimate.pdf.

Part II

Economic management of spectrum

4 Using auctions to assign spectrum

4.1 Introduction

In economics, auctions form a part of a field of study more generally known as *mechanism design.* As Hurwicz and Reiter [1, p. 30] explain, "in a design problem, the goal function is the main 'given,' while the mechanism is the unknown. Therefore, the design problem is the 'inverse' of traditional economic theory, which is typically devoted to the analysis of the performance of a given mechanism." In designing an auction the spectrum manager exercises a degree of discretion in pursuit of the chosen goals. This chapter discusses the properties of different types of auction design. Useful general accounts of auctions can be found in [2, 3]; on spectrum auction design see especially [4].

While the use of auctions to assign licenses is now the default position for most major spectrum award processes worldwide, before 1989 this was not the case. To understand why auctions have become so prevalent, it is useful to examine deficiencies in other processes. This is done in Section 4.3.

4.2 Some types and effects of auctions

An item up for auction can be described as having either *private-value* characteristics or *common-value* characteristics. This distinction can matter fundamentally.

In a private-value auction each bidder values an item by reference to the value it confers through consumption. For example, in deciding how much to bid for a family heirloom, a person is likely to be guided by sentimental rather than financial considerations. There is no reason why one person's valuation should be the same as another's.

By contrast, a common-value auction typically involves items that are not exclusively or chiefly used for final consumption. The key characteristic of an item in a common-value auction is that all bidders would agree on the valuation if they possessed the same objective information about the item. A good example would be a tract of land made available for drilling rights to extract oil. If all parties interested in the tract were given the same geophysical data, were to use the same extraction techniques, and estimated the same futures prices for oil, then they would each have the same belief about the valuation – in other words the valuation would be common to all bidders [5].

In practice, it is unlikely that all bidders in a common-value auction will be in possession of the same information. This leads to a problem which is key for auction design – the winner's curse: see Box 4-1.

> **Box 4-1.** Auctions and the winner's curse
>
> Consider the following hypothetical example. Assume there are three interested parties bidding for a common-value asset, with valuations of $100 million, $120 million, and $140 million. The difference in valuations arises because of differences in assumptions about the future development of the sector.
>
> The average valuation among the interested parties is $120 million. We could suppose that if someone pooled all the information held by the interested parties they would conclude that this was the expected value of the oil tract. Intuitively this estimate of the valuation is more likely to be correct as it uses more information – or, in the language of statistics, the valuation estimate is more efficient the larger the sample size. Suppose that the government is auctioning the asset via a sealed-bid mechanism that grants it to the highest bidder, at a price equal to that party's bid. It is reasonable to speculate that the party regarding the tract as worth $140 million would bid the highest amount, but might hope to get away with, say, a bid of $130 million.
>
> If the true valuation is $120 million, the winner has overbid. And this is likely systematically to be the case as the more optimistic a bidding firm's valuation, the more likely it is to win the auction. On this basis the winning firm, instead of celebrating its victory, might sensibly worry that it had paid too much.
>
> This is an example of a phenomenon known as the *winner's curse*, which means that the winner, which may not be the asset's most efficient user, risks finding itself in financial difficulties. Sophisticated bidders, concerned about falling victim to the curse, would shade their bids down to avoid it.
>
> The winner's curse is thus a danger for auction designers. How can it be mitigated? One means is to build into the auction process a degree of information transfer among bidders – carefully designed, of course, to avoid collusion. We discuss this further in section 4.3 below.

In the real world, auctions tend to have elements of both private and common values and in this respect radio spectrum is no different. Radio spectrum is not a final-consumption good and its use within commercial spheres (particularly in telecommunications) gives rise to common values. As entities making use of radio spectrum differ in terms of their outputs, goals, or initial endowments, their valuations are likely to have a private-value component. For example, mobile telephone companies competing for the same customers might have different strategies concerning the pricing of their mobile broadband offerings. Also, on the cost side the firms will likely exhibit different network topologies, have different vintages of equipment and possess different skillsets in their labour forces. These factors will lead them to valuations of radio spectrum which have much in common but are not identical.

4.3 Designing mechanisms to award spectrum licenses

A spectrum manager seeking to award licenses in a frequency band will initially face the policy challenge of how best to organize the granting of licenses. Although auctions are usually the default starting position today, this has not always been the case. We therefore consider a number of mechanisms in order. They are listed below:

1. first-come, first-served;
2. lotteries;
3. comparative selection (also known as beauty contests);
4. auctions.

4.3.1 Mechanism 1: first-come, first-served

In this case the spectrum manager awards licenses for use when an application is submitted. This approach performs satisfactorily when demand lies significantly below the number of licenses (or frequency) available and such a state is expected to persist.[1] In other words, this approach may be suitable when the opportunity cost of marginal radio frequency is zero, because it has no other worthwhile use. However, for situations where parties compete for frequency bands this approach is unsatisfactory and can give rise to corruption.[2] The latter problem may be so acute that it undermines service provision and results in much economic inefficiency. Where demand proves to be large relative to available spectrum, alternative mechanisms are required to award licenses more effectively. In the United States in the 1980s policy shifted from first-come, first-served to lotteries as a way of mitigating corruption and dealing with competing claims for licenses.

4.3.2 Mechanism 2: lotteries

In a lottery, a license is awarded by a random draw which each interested applicant has an equal chance of winning. As a consequence there is no feature in the process which tends to direct the license in the first instance to the firm which has the highest valuation, which in many cases will be the firm which can use it most efficiently.

However, if resale of the license is allowed, use of a lottery is likely to invite speculators – parties with relatively low valuations who seek to win the license solely in order to resell it at a higher price by running an auction of their own. This may promote efficiency, but the revenues from the auction accrue to the speculator, not to the issuer of the license.

---

[1] For example, suppose there are five possible licenses to be awarded in a frequency band and there are three interested parties. These potential licensees can apply to hold a license at any time. This mechanism may well be the most effective for this scenario.

[2] Corruption was associated with a 2G license award process in India in 2008 which was based on a first-come, first-served mechanism. In this case certain government officials brought forward the application date for the licenses and informed only certain parties about this change, resulting in favoritism and, by implication, rent extraction by those responsible for the award process. See the 2011 Report of the Comptroller and Auditor General of India for the year ended March 2010 [6].

This process also results in delay to service provision and consequent loss in consumer well-being, a feature that beset many award processes in the United States in the 1980s.

Commenting on the use of lotteries in the USA in the 1980s, the regulator (the FCC) wrote in 1997,

"Application mills" sprang up to assist almost 400,000 different firms claiming to be spectrum "providers" in their efforts to win a cellular license, and a broad range of spectrum speculators participated in and won lotteries in cellular, Specialized Mobile Radio ("SMR") and other services. Many license winners, with no intention of providing service to the public, were now eager to trade their license rights for windfall profits, and a secondary market in FCC licenses emerged. Even when lotteries themselves could be conducted quickly, it took years for secondary markets to reassign licenses to the parties that valued them the most and to aggregate these licenses efficiently. Delay in service to the public was often the result.[3]

Rather than grant licenses randomly, an alternative mechanism is intended to elicit privately held bidder valuations via a competitive process. Two mechanisms are discussed: comparative selection (or hearings) and auctions. We now discuss each of these.

4.3.3 Mechanism 3: comparative selection

This approach works in practice by each bidding firm submitting a detailed document outlining its intended use of and service plans for the radio spectrum, often followed by an oral hearing. Thus the consideration affecting the assignment decision is not the price offered for the license, as in an auction, but the characteristics of the service to be provided.[4] In broadcasting this might involve the range and quality of programs to be offered. In mobile communications it might include the geographical area over which service will be provided. This may make sense if the goal is to achieve the "best-quality" outcome in a context where license revenues are a matter of indifference, but it does carry the risk that the result might be that the level of quality offered in the winning application is unnecessarily high or unsatisfactorily low.

However, the main problem with comparative selection concerns the incentives that bidders have to be truthful when compiling their applications. Bidders may recognize that overpromising will increase their prospect of success, and act accordingly by offering unrealistic levels of performance. In principle this could be deterred by an appropriate system of penalties for broken promises. But these will only come to light some time into the operation of the license, when the imposition of financial penalties may cause performance to deteriorate even more, and cancellation of the license may lead to interruption of the service.

Comparative selection usually relies ultimately on judgments of quality, however detailed the scoring system which is developed to codify the judgments. This creates the danger of corruption or favoritism. This feature may also open the door to legal challenges, which result in delay.

---

[3] See http://wireless.fcc.gov/auctions/data/papersAndStudies/fc970353.pdf, 7.
[4] It is possible to run a comparative selection process in which proposals include, as well as the characteristics of the proposed service, a monetary sum offered. These are sometimes called "menu auctions" [7].

Comparative-selection hearings may be appropriate in some contexts, but the problems they face are considerable. In the sector where most competition for the spectrum licenses has recently occurred – mobile communications – comparative selection has become the exception rather than the rule. Operators often favor it, perhaps hoping to avoid substantial outlays on acquiring licenses. But the absence of license revenues is one reason for which governments reject comparative selection.

4.3.4 Mechanism 4: auctions

The obvious alternative to a comparative-selection process is an auction, in which firms make competing monetary bids for a license granting access to spectrum.[5] As noted above, the valuations made by bidders in a spectrum auction will be based on expectations of market developments broadly held in common and on factors specific to the bidder, such as the detailed topology of its existing radio access network. They will also be subject to error.

For an efficient allocation, the license should go to those with the highest overall valuations (minus errors), and much auction design is directed to this goal. Auction formats vary. Some require a single bid to be placed whereas other auction formats allow for bids to ascend over time, or possibly descend over time from an initial high value announced by an auctioneer. The different auction formats are reviewed in greater detail below.

We discussed in Box 4-1 the way in which the winner's curse can be an obstacle to the efficient assignment of spectrum by auction: the problem is that the highest bidder may not be the most efficient user of the license, but the bidder with the most optimistic expectations of its profitability – the highest positive error. Consider an auction design in which all bidders are invited to submit a single bid and are informed that the highest bid received will be awarded the license and pay the amount bid (this is known as the *first-price sealed-bid* auction). Each valuation will be made up of the components noted above, but crucially each bidder has to rely on its own information: there is no sharing or pooling of information among bidders.

Suppose instead that the auction featured an ascending price, as is typical in *open-outcry ascending* auctions familiar in the art world or in online auctions such as eBay. The auctioneer opens with a reserve price, and bidders drop out as that price is raised. The exit of bidders is important as it provides information for the valuations of survivors, who can draw inferences from the successive revelation of the upper limits on other bidders' valuations.

Other auction designs utilize the parties' information and judgements in different ways. For example, a *second-price sealed-bid* auction proceeds in a manner similar to the first-

---

[5] Experience suggests that the timing of the monetary payment is significant. Requiring it to be made up front limits the auction to those with adequate resources. But allowing payments in instalments creates conditions in which the licensee can exit if things go less well than expected, leading to disruption of service. An irrevocable up-front payment also turns expenditure on access to spectrum into a sunk cost, which firms take into account in setting their bid limits, but should not take into account when setting their service charges when the license payment has been made; see section 4.4 below.

price sealed-bid auction noted above, except that the winning bidder pays not the price it bid, but the price bid by the second-highest bidder. In other words, bidders go into the auction knowing that their chance of winning the auction depends on their own bids, but that the price they have to pay depends on the (lower) amount bid by another party.[6]

In general, where common-value factors play a role in valuations, price discovery should be allowed to work so as to facilitate indirect pooling of participants' estimates of them.

In a well-designed auction which is won by the highest bidder, there should be no ambiguity over who is the winner, and hence no (or much reduced) scope for corruption or favoritism. Auctions are, however, subject to bid-rigging, or collusive agreements by bidders not to compete with one another. For example, in an auction in which 80 MHz of spectrum is available, the four mobile operators seeking spectrum might reach an understanding that each will bid for only 20 MHz. As a result, the price will be the reserve price set by the regulator, and the operators will have enhanced their profits. We discuss in section 5.4 how this form of collusion can be combated.

Finally, spectrum auctions have redistributive consequences. We saw above that a lottery randomly distributed a scarce resource among individuals, and that those individuals could capture the associated scarcity rents either by using the resource themselves, or by organizing their own private auction to sell it on. In a highly contested comparative-selection process, competitors dissipate any excess profits by increasing the quality characteristics of their offering; in less competitive processes, the winner can appropriate some of the profits for itself.

In a well-designed auction, the revenues go to the state. In other words, the finance ministry appropriates the scarcity rents associated with spectrum, usually for general government expenditure purposes, occasionally to meet objectives associated with the communications sector. This is a form of revenue-raising which is particularly efficient, because it falls on the scarcity value of a natural resource in fixed supply, rather than falling (as income taxation does) on other productive inputs such as labour, the supply of which can be affected by the taxes the government raises. However, this efficiency property only applies if the government does not succumb to the temptation to act as a monopolist itself and deliberately restrict the supply of spectrum to increase auction revenues. In other words, auctions have the capacity to distribute spectrum efficiently, provided that both sides of the market (government and bidders) do not abuse market power.

4.4 The spectrum auction process

As noted above, until the late 1980s administrators assigned licenses using processes that included awards on a first-come, first-served basis; lotteries; and comparative selection. In the early 1990s a few administrators chose to auction spectrum rights, and, following the large revenues raised in auctions for mobile telecommunications spectrum rights in the United States in the mid-1990s, interest in using auctions to assign

---

[6] As we see below, the second-price sealed-bid format also makes it easier for bidders to decide how to bid, given their valuations.

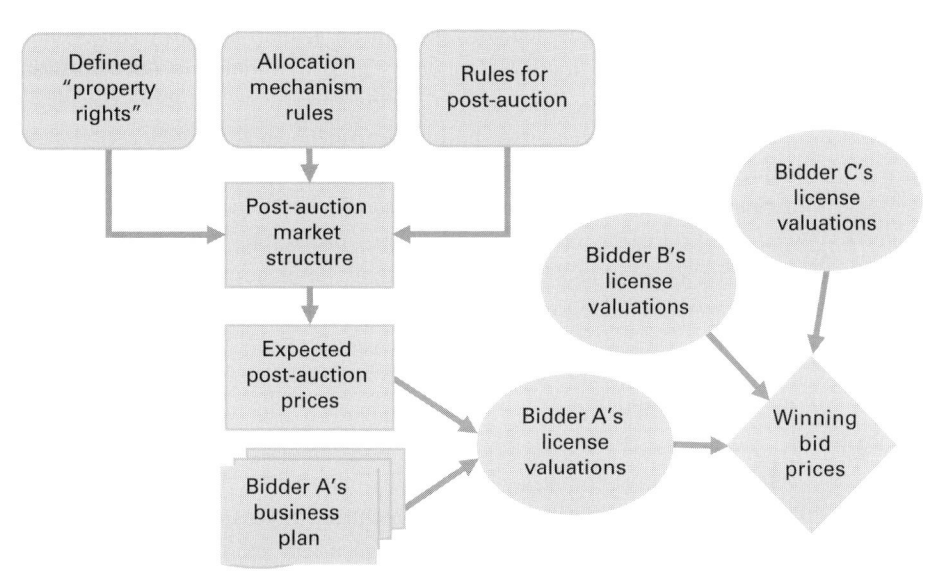

Figure 4-1. Factors influencing maximum bid values in a spectrum auction.

frequency rights increased markedly around the world. Advocates of auctions, beginning with Ronald Coase in 1959, have long argued strongly that the outcomes in well-designed auctions are better for society than administrative procedures.[7] It is widely argued that the superiority of auctions stems from their objectivity and transparency.

An argument commonly leveled against auctions is that bid prices necessarily result in consumers paying more for services reliant upon radio spectrum. In well-designed auctions, bidders are required to pay for spectrum up front. This means that successful bidders will treat the cost of spectrum as sunk. It is well known from economic theory that sunk costs do not influence market prices. The amount which bidders in a spectrum auction are prepared to pay for a license depends on their *expectations* of what service price will be, but once the auction price has been paid up front it is a sunk cost which does not affect service prices; those prices are determined by demand, by forward-looking costs, and by the degree of competition in the market where the service is sold. An empirical study supports this view. After cellular spectrum was auctioned in different regions in the US, the prices charged for mobile services did not vary as widely as prices for radio spectrum, and there was no statistically significant correlation between auction fees and the prices paid by consumers for mobile services [8]. Subsequent work has confirmed this [9].

The factors which influence bid values in an auction are illustrated diagrammatically in Figure 4-1. The key factors are the variables which determine the shape of the market structure in the post-auction phase. For example, the number of competitors in a market will be influenced by the allocation mechanism rules employed in an auction (such as the number of licenses), by the nature of the property rights offered (such as the duration of the licenses to be auctioned), and by post-auction rules (such as whether trading of

---

[7] See Box 3-1.

spectrum is permitted). Knowledge of these factors and expectations about the post-auction competitive market structure is used by bidders to develop business models which then determine their maximum bid value.

Maximum bid values may be interdependent, in that Bidder 1's valuation may be influenced by Bidder 2's valuation, and depending upon allocation rules these could change during an auction. Not surprisingly, in all auctions the valuations of bidders determine the outcome, and in a good spectrum auction – that is, one that is well designed – spectrum should be assigned to those bidders holding the highest valuations. As noted below, the degree to which bidders offer their highest valuation depends on the auction format chosen.

4.5   Auction theory

Radio spectrum auctions have historically featured a seller and potentially many buyers. The seller is usually an agent of government and buyers are typically commercial entities (e.g. telecommunication companies, broadcasters, etc.). The spectrum made available for auction is often spectrum that was formerly used for other purposes. For example, following the digital switch-over in terrestrial broadcasting, much spectrum in recent years has been refarmed from broadcasting use to telecommunications. In one case, the United States, it is planned that this transfer should be effected in a novel way, called an incentive auction [10]. It is described in more detail in section 4.9 below.

In general, there are two main forms of spectrum auction: (i) interactive or open auctions and (ii) sealed-bid auctions. Interactive auctions occur where bidders interact with the auctioneer through a price-discovery process that typically features either ascending or descending bids. An example of a popular interactive auction is the English auction in which bidders submit bids in an ascending manner until the number of bidders remaining equals the number of objects for sale. Another interactive auction is the Dutch auction, where the auctioneer starts with a high price and reduces this over time until a bidder accepts the posted price.

Sealed-bid auctions occur where bidders submit their bid once only in a sealed envelope (or more likely in practice via electronic means). The amount a successful bidder pays in a sealed-bid auction depends on whether it is a first- or second-price auction. In a first-price auction, successful bidders are required to pay the amount indicated on their bid, whereas in a second-price auction the successful bidder pays an amount equal to the second-highest bid submitted. Vickrey [11] presented the first theoretical assessment of these different auction forms.

Auction theory can be used to assess how "efficient" the different auction formats are, where efficiency is measured by the degree to which the auction assigns the lots to those holding the highest valuations of them. The theory of auctions can also shed light on optimal bidding behavior in auctions. Under special conditions the amount of revenue an auctioneer should expect to receive in an auction has been shown by auction theorists to be invariant across a wide range of auction formats [2, 11]. This result, known as the *revenue equivalence theorem*, has a significant bearing upon practical auction design and spectrum management policy – see Box 4-2.

Box 4-2. The revenue equivalence theorem (RET)

Probably the most famous result in auction theory is the revenue equivalence theorem, originally expressed by William Vickrey in 1961 [11]. According to this theorem, on certain conditions the auctioneer can expect the same revenues from any of a number of standard auction designs. This applies whether the object being auctioned has a purely private value (based solely on the preferences of the bidder, for example her aesthetic tastes), or whether it has a common, but not identical, value to a number of bidders (as would be the case of an input into a production process such as spectrum).

These designs include

1. a first-price sealed-bid auction, in which the object goes to the highest bidder, at the price that person bids;
2. a second-price sealed-bid auction, in which the object goes to the highest bidder, at the price bid by the second-highest bidder.

At first sight this looks paradoxical, until it is recognized that what potential buyers bid depends on the rules of the auction. They will be emboldened to bid more if they pay the second price.

Other familiar auction designs yield the same revenue:

3. an ascending auction, in which bidders successively cry out their bids until only one remains;
4. a descending auction, in which the auctioneer lowers the price until the object is bought.

So do some less familiar ones, including

5. an all-pay auction, in which the object goes to the highest bidder, but all bidders pay their bid.

The conditions on which the result holds are partly technical, such as that each bidder has a value for the object drawn from a specified distribution, and that any bidder with the lowest feasible valuation expects to pay nothing. Bidders also have to employ an "equilibrium bidding strategy."

Another required assumption is more intuitive: the bidders have to be risk-neutral. That is, they have to be indifferent between a certain gain of $$X$ and equi-probable outcomes of a gain of $0 and of $2X$. We can see how relaxing this assumption would undermine the RET. A bidder in a second-price sealed-bid auction bids her valuation whatever her attitude to risk is.[8] But the same risk-averse bidder in a first-price auction would, if averse to the risk of losing the auction, shade her bid upwards because she is willing to sacrifice an amount of expected surplus to make it more

---

[8] This can be seen as follows: if a firm bids more than its true value, it may be left to pay an amount which exceeds that value, if the second price is very close to its own bid; if it bids less than its true value, another bidder may snatch the prize at a price less than our bidder's true valuation.

certain that she will win the auction. Thus the form of risk aversion makes the expected revenue in the two auctions different, contradicting the revenue equivalence result.

This and other considerations have led to the paradoxical result that, because the RET depends on unlikely assumptions, it has focussed more, rather than less, attention on auction design, as the rest of this chapter and the next show. But the RET has greatly helped to identify the factors which affect revenue and other outcomes in an auction and to provide guidance on auction design.

In general, the key factors which influence expected revenues in an auction are related to the risk characteristics of the buyers and seller, and to the amount and type of information held by bidders about the value of the objects. In real-world auctions, there is likely to be much heterogeneity across bidders (e.g. incumbent mobile operators versus entrants) and this is likely to have a bearing on auction outcomes and hence on the auction design chosen.

In addition, the intensity of competition, for which the number of bidders is a proxy, generally has an impact on auction outcomes, and thus on the choice of auction rules. Auction theorists have shown for a wide class of auctions that a simple ascending auction with similarly placed bidders (i.e. bidders whose holding of information is symmetric) will generate a better outcome the more bidders there are [12]. This result is of special interest in the context of auctioning radio spectrum, as the number of bidders in high-stakes auctions has been in many cases relatively small. Hence it may be worthwhile to seek to attract extra bidders to participate.[9]

Auction theory is also helpful in deciding other aspects of the auction. These include the number of licenses, or more generally how best to package spectrum rights; how to determine the geographic scope of licenses; the eligibility criteria for bidders; and the precise auction rules. We return to these questions below in this and the next chapter.

4.6 Auction objectives

The format of an auction depends on a number of factors, the most significant being the objectives of the owner or custodian of the resource, which is usually government. If the custodian wishes to maximize revenue from the sale of spectrum, this may entail a different auction design than for a case where effective competition is the primary objective. Other important factors will be the amount of spectrum available, the nature of demand-side interest in the spectrum, the geographic area covered, and administrative factors. In many cases auctions for radio spectrum have been designed with efficient

---

[9] In the UK 3G auction the merchant bank N. M. Rothschild was offered a contract containing an achievement fee which created an incentive for it to attract more bidders into the auction. If seven bidders participated it would be paid £300,000, if nine bidders participated it would obtain £500,000 and if there were at least 11 bidders the bank would receive £700,000. In the end there were 13 bidders.

assignment in mind; that is, the auction is designed to assign spectrum to the highest-value use. However, governments have also recognized that high-value spectrum can be a useful way to gain revenue.

In the United States the Federal Communications Commission (FCC) and the National Telecommunications and Information Administration (NTIA) share responsibility for managing spectrum. Section 309(j) of the Communications Act authorizes the FCC to use auctions to promote efficient and intensive spectrum use as well as to promote the development and rapid deployment of new technologies, products, and services for the benefit of the public, including those residing in rural areas, without administrative or judicial delay. The Act also requires the FCC to administer auctions in order to promote economic opportunity and competition. Other objectives in the US include advancing spectrum reform by developing and implementing market-oriented allocation and assignment-reform policies, and conducting effective and timely licensing activities that encourage efficient use of the spectrum [13]. However, some writers have questioned whether the objective is efficiency and suggested instead that auctions are primarily a device for raising revenues.

The approach in Europe with regard to spectrum assignments has been to ensure harmonized conditions for the availability and efficient use of radio spectrum [14]. Licensing rules in the European Union require that where the demand for radio frequencies in a specific range exceeds their availability, appropriate and transparent procedures should be followed for the assignment of such frequencies [15]. In practice this has led to endorsement of the assignment of valuable radio frequencies by auction. Spectrum regulators are expressly prohibited by the EU Authorisation Directive from limiting access to spectrum for reasons other than ensuring its efficient use (for example, to maximize revenues).[10]

In New Zealand, the first country to pioneer spectrum auctions, the government's policy objectives for spectrum encompass the promotion of competition, maximizing the value of spectrum to society, and satisfying increasing demand. When assigning spectrum rights the government is required to adopt a market-based assignment process which is competitively neutral and transparent.

Spectrum auctions have been used in many other countries and in most cases the objectives are similar to those outlined above and can be summarized as follows:

- **Efficiency:** licenses are awarded to those who value them the most; alternatively, they are awarded to those whose use of them contributes most to economic and social welfare.[11]
- **Revenue:** revenue is raised for the government.
- **Competition:** spectrum rights are issued in a way that helps promote effective competition in spectrum-using services.
- **Transparency:** the process of selection is accomplished without corruption and expeditiously.

---

[10] Authorization Directive 2002/20.EC, Article 5.
[11] The difference between these two formulations is examined in sections 11.6–7.

The challenge for spectrum agencies is to identify the appropriate auction formats for different types of radio spectrum use and different users that broadly achieve some or all of the above objectives. Below we discuss the most popular auction formats that have been applied by spectrum agencies around the world since 1989, as well as some formats that may be applied in the future.

4.7 Auction formats

In this section we describe seven auction designs, identify their pros and cons, and discuss some spectrum applications. It is followed by two further sections devoted to combinatorial clock auctions and to an "incentive" auction; both of these are important and recently developed formats for spectrum auctions.

Note that in the auctions described in this section, it is assumed that lots which may be complementary to one another, or substitutes, are bid for independently, in the following sense. Suppose that spectrum licenses are being auctioned separately in each of two regions of a country – north and south. A firm's business may depend crucially upon gaining a license in each. Yet if the auctions are separately conducted (even if simultaneously) that cannot be guaranteed; the firm might end up with a license in *either* north *or* south. This so-called "exposure" risk of coming out of the auction with the wrong portfolio will influence the firm's bidding strategy, in a way which may imperil the efficiency of the outcome and the revenue expected. This significant risk is avoided in the "combinatorial" approach, discussed in section 4.8, which allows a bidder to make offers conditional upon its getting the combination of outcomes it wants overall.

4.7.1 First-price sealed-bid auction

A first-price sealed-bid auction is a very simple design that is easy to implement and easily understood by bidders: the bidder who posts the highest bid wins the license, and pays the amount bid. Note that this auction design is equivalent to a Dutch auction in which the auctioneer brings prices down until the lot is sold.

Pros:
- simple,
- completes quickly,
- ideal for small packages of spectrum and where each bidder seeks one license in a given region,
- good for competition – new entrants are more likely to enter and compete against incumbents.

Cons:
- Because there is no price-discovery process, bidders receive no information about others' valuations in the auction process. This may lead to the operation of the winner's curse. Alternatively, bidders may shave bids excessively to avoid it.

- Bidders winning nearly identical objects may pay widely divergent amounts; this can cause presentational problems for managers or politicians.
- Where there are many items to be auctioned and these comprise either substitutes and/or complements, this can give rise to an exposure problem (i.e. a bidder ends up with the wrong combination of licenses).
- Reserve price needs to be carefully assessed.

There have been many spectrum auctions using the first-price sealed-bid format. It was first used in New Zealand in October 1991 for FM, UHF, AM, and DMS frequencies for local licenses throughout the country.

In April 2006 the UK spectrum regulator Ofcom used this format to auction low-power services in the 1,781.7–1785 MHz spectrum band, paired with 1,876.7–1,880 MHz (the GSM/DECT guard bands). These bands can be used for a range of applications, such as private GSM networks in office buildings or campuses, though formally the auction rules stated that the spectrum was to be awarded on a technology- and application-neutral basis. Ofcom used an innovative approach by allowing bidders to submit different bid values depending upon the number of licenses that might finally be offered. Ofcom stated that it would issue a minimum of seven and a maximum of twelve licenses, and that the actual number of licenses would be determined by the auction process. This design feature allowed the market to determine how many operators should offer a service, rather than having the spectrum manager choose what it believed was best for customers.

In the auction fourteen bids were submitted. One bidder paid a little over £1.5 million for a license, whereas another paid £50,110.[12] The auction determined that 12 licenses would be issued, as the bids submitted conditional on 12 licenses being issued raised the highest revenue for Ofcom.

4.7.2 Second-price sealed-bid auction

A second-price sealed-bid auction is equivalent to an English auction and is also known as a *Vickrey* auction, in which bidders post sealed bids and the highest bidder pays the price offered by the second-highest bidder.

Pros:
- encourages bidders to reveal true valuations,
- ideal for small packages of spectrum and where each bidder will receive one license in a given region,
- completes quickly.

Cons:
- Where there is a large difference between the successful bidder's offer (the "first price") and the "second price" that bidder actually pays, it may be politically awkward to explain.

---

[12] The reserve price was £50,000.

Table 4-1. Result of the UHF auctions in New Zealand in 1989

	Near nationwide UHF TV frequency rights		
	Successful bidder	Price paid	Highest bid
Lot 1	Sky Network Television	NZ$401,000	NZ$2,371,000
Lot 2	Sky Network Television	NZ$401,000	NZ$2,273,000
Lot 3	Sky Network Television	NZ$401,000	NZ$2,273,000
Lot 4	Broadcast Communications Ltd	NZ$200,000	NZ$255,124
Lot 5	Sky Network Television	NZ$401,000	NZ$1,121,000
Lot 6	Totalisator Agency Board	NZ$100,000	NZ$401,000
Lot 7	United Christian Broadcast	NZ$401,000	NZ$685,200

- There is no price-discovery process in the auction, but the amount paid depends upon another bidder's valuation (which provides some protection from the winner's curse).
- Where there are many items to be auctioned and these comprise either substitutes and/or complements, this can give rise to an exposure problem (i.e. a bidder ends up with the wrong combination of licenses).

The first spectrum auction held in the world was a second-price sealed-bid auction for UHF frequencies in New Zealand in December 1989. This was followed by two more second-price sealed-bid auctions in 1990 in New Zealand. In the first auction the UHF frequencies result is shown in Table 4-1. One bidder, the Totalisator Agency, submitted bids of NZ$401,000 for each of six licenses. These bids determined the price of the five lots where the Agency bid second. In another lot which it won, the Agency paid the lower second price of NZ$100,000.

As this auction comprised national and regional frequency rights where the lots offered were either substitutes or complements, it is doubtful whether efficiency was achieved. This is because the auctions were independent, which meant that it was difficult for bidders to manage the risk of winning too many or too few licenses. Leading auction specialist Paul Milgrom has stated that it is likely that this led to an inefficient outcome [3]. A combinatorial auction, discussed below, could solve this problem.

New Zealand abandoned second-price sealed-bid auctions in 1990 and proceeded to use first-price sealed bids. This was largely in response to criticism leveled against the government for allowing large corporations to buy spectrum at knock-down prices, despite revealing in their bids that they would pay much higher prices. In one instance a bidder submitted a bid of $100,000 and paid $6 for a license, and in other cases, due to the absence of a reserve price, licenses were given away because there was no second bid [16]. This criticism is, of course, misplaced, as there is no guarantee that the firm which bid $100,000 in a second-price auction would have bid the same amount in a first-price auction (see Box 4-2 on the revenue equivalence theorem, or RET).

4.7.3 Simultaneous multi-round auctions

The simultaneous multi-round auction, or SMRA, is a format in which bidders submit bids round by round, successively bidding up prices until a round is reached in which no new bids are received. The auctioneer stipulates a minimum bid increment and requires that bidders participate if they are active (this is known as the Milgrom–Wilson activity rule). Activity rules are intended to prevent the "snake-in-the-grass" strategy, where a bidder may misrepresent early-round bids by understating demand. Bidder identities can be kept confidential during the auction, to help eliminate collusion and signaling (this was a feature of the 3G auction in Hong Kong in 2001). The winners in an SMRA should pay almost the same as in a second-price sealed-bid auction. The amount paid will on average be slightly higher due to the bid-increment requirement.

Pros:
- can deal with multiple lots,
- different prices paid will reflect differences in values between objects in auction,
- price discovery – especially significant where bidders seek to buy packages of licenses.

Cons:
- demand reduction – "large" bidders may understate demand early on,
- exposure problem – if some spectrum licenses are complementary a bidder may end up paying too much for some licenses if the prices of complementary licenses become too high in subsequent rounds,
- bidding strategies can be complex,
- vulnerable to collusion when competition is weak.

At the start of an SMRA, bidders are invited to submit bids on up to k spectrum blocks. For example, an SMRA may feature five different areas each having two spectrum blocks, making a total of 10 spectrum blocks, where a bidder can at most acquire one license in each area. In this example a bidder may be permitted to place bids on up to five spectrum blocks in the five different areas (hence $k = 5$). Suppose there are five bidders A, B, C, D and E.

Assume at the end of the first round that bidders A and C have submitted the highest bids on the two spectrum blocks in each of the five areas. In the next round, bidders B, D and E are invited to bid (they are the eligible bidders), while bidders A and C will be ineligible to bid. Should one of the other bidders submit a higher bid than A or C, the auction proceeds to the next round, with A and/or C again eligible to bid. The SMRA continues until a round is reached in which no new bids are submitted.

After each round in an SMRA closes, the bids in that round are disclosed. Auction specialists argue that this price-discovery process is one of the key strengths of the format. Because there are many related licenses in play (usually licenses in many different areas – though they could be different types of frequency package, such as paired and unpaired), the simultaneous sale of spectrum blocks and the ascending bid process aid price discovery and enable bidders to try to assemble desirable packages of spectrum.[13]

The SMRA format has assisted the price-discovery process in large-scale auctions (those auctions involving many bidders and/or many licenses) for awarding spectrum rights in the United States, for which it was principally designed.[14] The SMRA format was first deployed in the US in July 1994 for 10 nationwide narrowband PCS (personal communications services: mobile voice and data) licenses. The auction lasted five days and ended at round 47. Twenty-nine bidders participated and at the end of the auction six bidders won ten licenses, raising $617 million.

Following the successful application of the SMRA format for nationwide licenses, the FCC then organized a series of SMRAs for regional and local narrowband PCS licenses. In these auctions many bidders were interested in acquiring several licenses across different areas and the SMRA price-discovery property was critical to the success of the auctions.

In 1995, two blocks (A and B) of regional narrowband PCS spectrum generated net bid values of just over $7 billion, and during 1995–1996 the local narrowband PCS C Block auction raised over $10 billion in net bid values. The C Block auction lasted 184 rounds over the course of 84 bidding days. Some 255 bidders qualified to participate in the C Block auction. At the end of the auction 89 bidders won 493 licenses. At the time the scale of the auction was unprecedented and the sums raised were the largest ever seen for radio spectrum. These auctions were conducted electronically, which meant that the pace of the auction could proceed relatively quickly.

The success of the SMRA format in the US PCS auctions generated tremendous interest among radio spectrum managers around the world in auctions generally and in the SMRA format in particular. To date, SMRA auctions have been used for many different types of radio spectrum globally and over $200 billion has been raised using this format. The format has only recently been supplanted by a combinatorial format described in section 4.8 below.

New Zealand applied the SMRA format to the assignment of local FM radio spectrum licenses in October 1995 and September 1996, using a fax-based bidding process. From 1998 onwards the New Zealand authorities used Internet-based bidding procedures to manage the SMRAs – which enabled the auctions to be conducted more rapidly. Australia used the SMRA format in 1998 to assign mobile telephony spectrum, yielding revenues on a per-capita basis much higher than those raised in the US PCS auctions. In 1999 Canada applied the SMRA format for spectrum in the 24 GHz and 38 GHz bands.

In the year 2000 the SMRA format was used to assign licenses for 3G spectrum in a number of European countries. Unlike the application of the SMRA format in the US, the 3G license assignments in Europe were relatively small-scale in terms of license numbers, involving usually no more than six nationwide licenses, and the number of qualified bidders in most cases was also relatively small. Table 4-2 provides data on the

---

[13] The combinatorial auction design described in section 4.8 below offers a more certain way of constructing packages of spectrum holdings.
[14] Spectrum assignment in the US is often undertaken on a regional and local level and many hundreds of licenses can be offered simultaneously.

Table 4-2. 3G SMRA auctions in Europe, 2000–2001

Country	Date	Number of bidders	Number of licenses	Comments
United Kingdom	April 2000	4 incumbents 9 new entrants	5	Auction extended over 52 days and 150 rounds
Netherlands	July 2000	5 incumbents 1 new entrant	5	Auction extended over 14 days and 305 rounds
Germany	August 2000	4 incumbents 3 new entrants	6	Auction extended over 19 days and 173 rounds
Italy	October 2000	4 incumbents 2 new entrants	5	Auction extended over 2 days and 11 rounds
Austria	November 2000	6 bidders	6	Auction extended over 2 days and 14 rounds
Switzerland	December 2000	4 bidders	4	Auction lasted 1 day
Belgium	March 2001	3 incumbents	4	Auction extended over 2 days and 14 rounds

number of licenses available in the main European 3G auctions using the SMRA format held over the period 2000–2001.

It is debatable whether the SMRA format added much value to the efficient assignment of spectrum in the European 3G auctions. Bidders in these auctions were not assembling packages of different licenses (except sequentially across different auctions in different countries), the number of licenses in each auction did not exceed six, the number of bidders was small, and the bidders were arguably well-informed sophisticated entities which had sufficient time to prepare and assess the value of the licenses. In these circumstances the SMRA was vulnerable to collusion, and it is unclear what would be gained from price discovery. A suitably designed second-price sealed-bid auction format would have been simpler, quicker, and arguably more effective at revealing bidders' valuations and less vulnerable to collusion. In other words, the process could have been much simplified.

The 3G auction in the UK in the year 2000 resulted in the largest revenue ever raised from a spectrum sale when measured on a per-capita basis. For each person in the UK, £375 ($694) was raised in revenue, which contrasts with the PCS C Block auction in the US, where revenue per person was much lower at a little over $38.[15]

The UK 3G auction offered paired spectrum, which is especially valued for mobile communications applications, and unpaired spectrum, which was valued less. Five national 3G licenses were made available by the spectrum agency in the UK, with

---

[15] The US population at the time of the auction was 262 million.

Table 4-3. Results of the UK 3G SAA

License	Winning bids £ billion	Price per MHz £ million	Successful bidder
A	4.385	125	TIW
B	5.964	199	Vodafone
C	4.030	161	BT Cellnet
D	4.004	160	one2one
E	4.095	164	Orange
Total	**£22.478**		

Table 4-4. Bidding process in the UK 3G auction

Bidder	Round	License	Bid value £ billion
Crescent	90	C	1.819
3G	90	A	2.001
Epsilon	94	C	2.072
Spectrum	95	D	2.100
OneTel	97	E	2.181
WorldCom	119	D	3.173
Telefónica	131	C	3.668
NTL	148	C	3.971

Source: NAO.

license A comprising 15 MHz paired and 5 MHz unpaired spectrum reserved exclusively for new entrants; license B was 15 MHz paired; and licenses C, D, and E were identical in size at 10 MHz paired and 5 MHz unpaired. Thirteen bidders competed in the auction, of which four were incumbent cellular phone network operators.

The result of the UK 3G auction is shown in Table 4-3. The auction ended after 150 rounds on April 27, having lasted seven weeks. The successful bidders were the four incumbent operators and a new entrant, TIW, which subsequently sold the A license to Hutchison Whampoa. On a per-MHz basis, the prices paid for licenses C, D, and E reflect the arbitrage property of SAA.

Table 4-4 illustrates the highest bids submitted by each of the bidders in the UK 3G auction. It can be seen that five bidders exited when the bid values were around the level of £2 billion per license, which occurred between rounds 90 and 97. After this the auction continued with eight bidders until round 131, when WorldCom exited the auction, and thereafter bid values increased to over £4 billion per license. The last unsuccessful bidder in the auction was NTL, which submitted its last bid, value £3.97 billion, in round 148 on license C. Earlier in the auction in round 127, NTL had

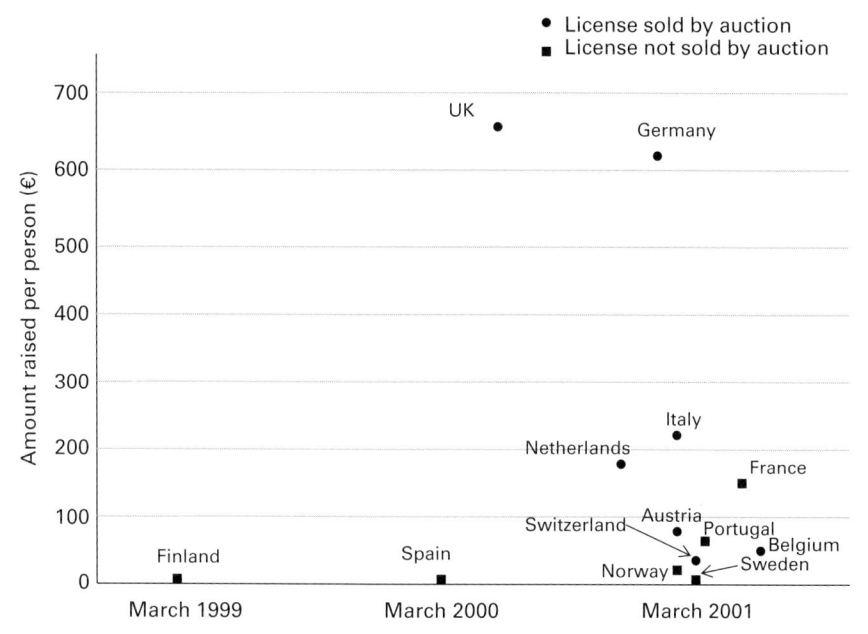

Figure 4-2. 3G auction bid values around Europe. Source: NAO.

submitted a higher bid of £4.28 billion on the larger license A, though this was equivalent to £143 million per megahertz versus £159 million per megahertz in its final bid for license C.

Similar SMRA auctions for 3G spectrum were held in a number of other European countries and the outcomes are illustrated in Figure 4-2. The revenues received vary substantially, with later auctions doing particularly badly. The auction held in Switzerland was a notable disaster, as the number of bidders ended up equal to the number of licenses, leading to weak competition and collusive behavior as bidders exited or formed larger coalitions prior to the start of the auction.

The comparatively low prices paid at 3G auctions that were held in late 2000 coincided with changing market sentiments at the time, reflected in the so-called "dotcom crash." In the UK, where revenues from the 3G SMRA auction had vastly exceeded expectations, only a few months later in November 2000 an SMRA for broadband fixed wireless access (BFWA) spectrum in the 28 GHz band was a notable failure in that few of the licenses were sold.

The UK BFWA auction was conducted on a regional basis with three licenses available in each of 14 regions, and each bidder was restricted to one license in each region. The auction featured substitutes (the three licenses in each region) and complements (licenses in the different regions) and an SMRA format was by far the most suitable format then available.

Only 16 licenses in eight regions were sold out of the possible 42 licenses. The total proceeds amounted to £38.16 million, compared to predicted proceeds of over £1 billion. In this auction, the reserve prices did not reflect the adverse market conditions in the telecommunications sector during the latter half of the year 2000. The reserve price was too high and there was no procedure for dealing with unsold licenses.

The setting of reserve prices is often an important feature of spectrum auctions, though it is not always necessary.[16] The purpose of a reserve price is to reflect the value of spectrum to "society" and to ensure that it is not transferred at knock-down prices below the true economic value.[17] Indeed, if reserve prices are set too low, this could encourage unhealthy tacit collusion as bidders may benefit significantly from an early conclusion to an auction. This may have been one reason contributing to the failure of the Swiss 3G auction in 2001, when the licenses were eventually assigned at values (assessed on a per-capita basis) significantly below those in comparable markets such as Germany and the UK.

On the other hand, if reserve prices are set too high this can reduce participation, which is bad for competition and the price-discovery process. In the UK BFWA auction the reserve price was based on data that were out of date by the time the auction occurred. The setting of reserve prices is challenging and requires careful judgment about market prospects. For spectrum administrators setting reserve prices, erring on the side of caution and being slightly pessimistic are probably advisable, though being overly pessimistic can give rise to embarrassments, as in the Swiss 3G auction.

The SMRA format was used to assign major spectrum in the United States in 2006. This auction was for advanced wireless services (AWS) or 3G. The AWS auction, like previous such auctions in the United States, illustrates that the SMRA format is very scalable and can cope with large numbers of bidders and licenses. The FCC auction of AWS spectrum licenses started on 9 August 2006 and ended on September 18 2006 in round 161. Of the 1,122 licenses offered, 104 bidders won 1,087 licenses, raising a total of $13.9 billion. Its successor in 2015, AWS-3, took over two months and yielded $45 billion from over 1,600 lots, equivalent to nearly $140 for every man, woman, and child in the USA. The auction was criticized for the large number of rounds required – 341 over 75 days. The suggestion has been made that it might be preferable, instead of auctioning each lot separately, to aggregate similar lots and conduct a clock auction for them, in the manner described next.

4.7.4 Ascending clock auction

In an ascending clock auction (also known as a *Japanese auction*, a *button auction* or an *Ausubel auction* [17]) bidders submit their demand in response to announced prices issued by the auctioneer. The auction continues until a round is reached in which aggregate demand is equal to the number of spectrum licenses on offer. This format is easy to apply when there are identical (or nearly identical) objects substitute spectrum

---

[16] The Dutch 3G auction did not feature reserve prices and three of the six bidders in the auction submitted opening bids of zero. The auction finally concluded after 305 rounds after one of the bidders (Telfort) sent another bidder (Versatel) a letter threatening a lawsuit if bidding continued.

[17] The economic value of spectrum can be proxied by its opportunity cost – a concept which is discussed in detail in section 3.5.4.1 and Chapter 7.

licenses and bidders are seeking one license, though it is adaptable to circumstances where bidders may seek multiple units (i.e. several licenses) and where items may differ somewhat. In many respects the ascending clock auction has very similar properties to an SMRA auction, with the added attraction that it is more likely to yield an efficient outcome.

Pros:
- generates an efficient result,
- works well for auctions with substitute packages,
- simple,
- transparent,
- auctioneer can determine pace of auction and combat collusive bid signaling,
- good for multi-unit auctions.

Cons:
- may require two auctions (sequential, not simultaneous) if the auctioneer overshoots the valuations of the bidders,
- may require sequential auctions where packages differ.

The world's first ascending clock auction for radio spectrum was successfully deployed in Nigeria in 2001. In this auction three GSM licenses, identical other than with respect to their position in the band, were offered and there were five bidders. This auction also incorporated a sealed-bid phase in the event of overshooting by the auctioneer.

4.7.5 Revenue-share auctions

Revenue-share auctions are variants of standard auction formats in which bidders submit royalty bids based on the share of revenue they will keep from the service provided by the spectrum.

Pros:
- said to share market risks between auctioneer and bidders,
- simple.

Cons:
- the royalty spectrum license fee forms part of a recurring cost base, which means that market prices incorporate it as a cost.

The most significant spectrum revenue-share auction to date occurred in Hong Kong in 2001 for 3G licenses. The auction was designed allowing participants to bid royalties with a reserve price of a 5% royalty on network revenues. The auction was a second-price auction, as the royalty figure was determined by the highest unsuccessful bidder. In practice, the number of bidders equalled the number of lots, meaning that the reserve royalty was the price paid.

4.7.6 Hybrid auctions

Hybrid auctions combine auctions with beauty contests (combinations of auction formats, which are also hybrids, are considered separately below). The process of comparative selection usually takes place at the pre-qualification stage to determine which bidders become eligible to participate in the auction.

Pros:
- agency has greater discretion over qualification of bidders.

Cons:
- is susceptible to the deficiencies of a purely comparative selection process.

This approach was adopted during the registration and pre-qualification stage of Hong Kong's 3G auction.

4.7.7 Sequential auctions

Sequential auctions are auctions where spectrum licenses are offered in sequence. For example, if a spectrum agency is offering several identical licenses, it might decide to offer license 1 in auction 1, followed by license 2 in auction 2, and so on.

Pros:
- can work with most auction formats,
- easy to administer.

Cons:
- encourages strategizing among bidders and may give rise to peculiar outcomes.

National and regional BFWA licenses were auctioned sequentially in Switzerland in March 2000. Two national licenses were auctioned, the first on March 8 and the other on March 9. Four bidders participated in the March 8 auction and the successful bidder, UPC, paid SFR 121 million, outbidding FirstMark Communications, which bid SFR 115 million.

In the auction held the next day, the three unsuccessful bidders from the day before participated. The outcome resulted in FirstMark Communications winning and paying SFR 134 million, some SFR 20 million more than it had bid the previous day. The outcome of the auction is consistent with the theoretical literature on auctions, which suggests that prices may drift upwards in sequential auctions if there is a common-value element.

Turkey held a sequential auction in the year 2000 for two mobile telephony spectrum licenses in the 1800 MHz band, employing a rule that the reserve price in the second auction would equal the price paid in the first auction. This design was flawed as it encouraged strategic excessive bidding in the first auction, which led to a high reserve price in the second auction such that no bidder was prepared to pay. Thus the successful bidder in the first auction (Is-TIM) strategically generated a more concentrated market structure as a result of the poor auction design. Furthermore, valuable spectrum was left idle and the government lost revenue.

4.8 Combinatorial clock auctions

Many spectrum awards feature multiple objects made available simultaneously. For example, a spectrum agency may have 120 MHz of bandwidth available in packages of 5 MHz blocks. If the agency knows the optimum amount of spectrum that should be held by the optimal number of licensees, then the award process is relatively straightforward and an SMRA auction can be implemented. For example, it might be decided that two licenses of 2 × 30 MHz paired should be made available.

In practice a spectrum agency is unlikely to be in such an omniscient position and those most knowledgeable about market requirements are likely to be the parties interested in acquiring spectrum. In the presence of this informational asymmetry, it would be preferable to allow the market participants to signal to the agency the appropriate assignment. Hence, rather than predetermine that two licenses, each being 2 × 30 MHz paired, be auctioned, an agency might instead make available a collection of packages each containing 2 × 5 MHz of paired spectrum via a clock aution.

The argument in favor of this approach is strengthened if the auction contains spectrum in different bands, in circumstances where different potential purchasers may have different abilities to substitute one band for another. For example, it may only make sense for a mobile operator without a sub-1 GHz holding to bid for more spectrum if it is able to complete its portfolio by acquiring a holding in the missing band, while to another operator with such a holding in the first place, a higher frequency such as 1,800 MHz or 2.3 GHz may be a good substitute for sub-1 GHz spectrum.

An agency which adopts the combinatorial approach allows bidders to make offers for combinations of spectrum, on the basis that they will either be granted the whole combination, or get nothing at all. This protects them from the risk of being stranded with auction wins only in some of the spectrum they need. And they can make a range of combination bids, on the basis that the auction manager will only award one of them.

Consider a very simple example [18]. Suppose there are two objects A and B in an auction, and two bidders 1 and 2. Suppose the valuations held by the two bidders are as set out in Table 4-5. If bidders submit bids revealing their valuations and the auctioneer determines winning bids subject to maximized value, the outcome of the auction would

Table 4-5. First example of combination values

Bidder	Valuation of packages		
	A	B	AB
1	7	2	9
2	5	0	8

Table 4-6. Second example of combination values

Bidder	Valuation of packages		
	A	B	AB
1	4	2	9
2	5	0	8

be that bidder 1 would win the auction. This outcome would occur both if the lots were sold in two separate auctions and if there was a combinatorial auction.

Suppose, on the other hand, that the valuations held by the two bidders are as set out in Table 4-6. In this second case, if each lot were auctioned separately, lot A would raise 5 from bidder 2 and lot B would raise 2 from bidder 1. However, this is not an efficient outcome as a combinatorial auction would raise 9 from bidder 1.

The mathematics underlying the combinatorial auction are complex [19], and it initially took some courage on the part of auction managers to adopt them. But they have now become quite widely used, particularly where spectrum is being offered in different bands, which are partial substitutes one for another. A combinatorial auction allows bidders to devise different portfolios of spectrum holdings, and respond to changes in their relative prices as the auction unfolds.

Because the winning bids usually comprise combinations of holdings, it is not possible unambiguously to establish after the auction at what price per megahertz each band has been sold. However, ways have been found to approximate these values.

The pros and cons of this auction design can be set out as follows:

Pros:
- can work with most auction formats, though a sealed-bid approach is much easier to implement;
- can result in substantial lowering of the exposure problem;
- suited to auctions with a relatively small number of licenses which are complementary.

Cons:
- complicated and may be difficult for bidders and observers to understand, which may undermine participation and revenue;
- complex to implement;
- requires detailed market understanding to assess what packages are most desirable;
- may require sequential implementation to reduce complexity of auction;
- some pricing and other transparency sacrificed when viewed from a public accountability perspective.

We take as an example the UK auction held in 2014 involving spectrum at 800 MHz, supplemented by some other spectrum in other bands.[18] This was a combinatorial clock auction (CCA). As the name suggests it starts with a clock auction, described in section

---

[18] This account relies heavily on Myers [20].

4.7.4 above. This is followed by a combinatorial stage, then there is an assignment stage which matches successful bidders to specific frequencies in the band.

The frequencies in the auction were:

A. 800 MHz band:
   1. 4 lots of 2 × 5 MHz, and
   2. 1 lot of 2 × 10 MHz with an obligation to provide 98% indoor coverage;
C. paired 2.6 GHz band: 14 lots of 2 × 5 MHz;
D. concurrent low-power use of paired 2.6 GHz; and
E. unpaired 2.6 GHz band: 9 lots of 5 MHz.

The first of the three stages of the CCA is the clock stage (see section 4.7.4 above). The clock stage of the UK CCA was a multi-round, simultaneous, ascending-package auction. In each round the auctioneer announces a price simultaneously on each lot category, and provides information on aggregate demand on each clock in the previous round. Each bidder can make at most one package bid in each round. If there is excess demand on any category, the price of that category increases in the next round. The clock stage ends when there is no excess demand in any category.

The clock stage of the auction assists in price discovery, allowing bidders to take account of complements through package bids, and substitute as relative prices change. They can learn from other (anonymous) bids, e.g. observing the extent of excess demand on each clock at different prices. The evolution of prices and demand can assist the bidder to focus on the most relevant parts of its valuation model and identify the most relevant packages it could win (which is especially important in the presence of budget constraints).

Second, after the clock stage has ended there is the supplementary bids round. This is a single round of sealed bids in which a bidder can submit many (mutually exclusive) package bids (although with some restrictions). This stage allows firms an opportunity to express their preferences in a more nuanced way. For example, they can express relative values for larger packages displaying synergies compared to smaller packages, which may have not been possible for the bidder in the clock stage because clock prices are expressed in a single per-unit price.

Finally, there is the assignment stage. This comprises a single round of sealed bids, which enable bidders to express preferences over the specific frequency locations within each category.

Auction prices for the winning packages are based on opportunity cost using a second-price rule, i.e. highest losing bids. The intention is to encourage truthful bidding and enhance auction efficiency as the auction price depends only on bids made by others, not the amount bid by the winning bidder.

This UK CCA produced the result shown in Table 4-7. As the amount paid as a result of the assignment stage was £27 million, total auction revenues were £2,368 million. A review of the auction by the UK's National Audit Office broadly endorsed the organization of the auction but made a number of recommendations for the future [21].

Table 4-7. Result of Ofcom 4G CCA after the principal stages

Licensee	Frequencies assigned	Base price
Everything Everywhere Limited	796–801 MHz and 837–842 MHz	£589 million
	2,535–2,570 MHz and 2,655–2,690 MHz	
Hutchison 3G UK Limited	791–796 MHz and 832–837 MHz	£225 million
Niche Spectrum Ventures Limited	2,520–2,535 MHz and 2,640–2,655 MHz	£186 million
	2,595–2,620 MHz	
Telefónica UK Limited	811–821 MHz and 852–862 MHz	£550 million
Vodafone Limited	801–811 MHz and 842–852 MHz	£791 million
	2,500–2,520 MHz and 2,620–2,640 MHz	
	2,570–2,595 MHz	

This UK auction was further complicated by the inclusion of a special procedure to retain within the market the fourth operator by giving it privileged access to a small quantity of key spectrum. This so-called "spectrum floor" is discussed in Box 5-1 below.

4.9 Incentive auctions

The auction formats described above are so-called one-sided auctions: the units for sale (paintings, wine, farm animals, spectrum licenses) are determined in advance, if in some cases with "reserve prices" below which they will not be sold. The auction process then ensues, usually employing one of the mechanisms discussed above.

It is possible, however, to envisage, and in some cases to observe, another type of process: a two-sided auction, with participation from both buyers and sellers. Consider the following stylized example: two types of spectrum user, broadcasters and mobile operators, are both able to use a single band (say the 600 or 700 MHz band) to provide their service. Some have homogeneous rights of access to the band, which they can sell; others may want to buy, if the price is right. All are assembled in a room by an auctioneer, who initially proposes a trial price. Those willing to sell megahertz at the specified frequency indicate the amount; those willing to buy at that price signal how much they want to buy. The auctioneer adds up supply and demand, raises the price if there is excess demand, and lowers it if there is excess supply. When a balancing or equilibrium price is found, the appropriate transactions take place.[19]

It is clear that periodic auctions of this kind represent a market-driven mechanism for refarming, quite different from the "command-and-control" procedure normally used, in which the regulator determines the allocation of the spectrum (that is, its use), even if a one-sided auction is used to assign the licenses. The procedure described is thus potentially better placed than the administrative alternative to achieve the efficient allocation of spectrum illustrated in Section 3.2, where marginal returns in both activities are equalized. It does so by capturing in the individual offers and bids the private knowledge held by broadcasters and mobile operators of the value to them of additional units of spectrum.[20]

The two-sided approach can build on several aspects of auction practice or proposals. A firm may only be willing to sell in one band if it can at the same time buy in another. This need for synchronicity was part of the inspiration for an influential two-sided auction proposal made by Kwerel and Williams in 2002 [22]. This envisaged a large auction, the "big bang," which all licensees would have an incentive to take part in, although they would not be under any obligation to sell. This would not only lead to the discovery of the worth of different bands, but also create conditions for synchronized buying and selling, making it easier for spectrum users to maintain continuity of service for their customers.

The degree of certainty available to buyers and sellers would be further enhanced if the combinatorial, or combinatorial clock, approach described above were adopted. This would enable buyers to make bids and offers defined in terms of combinations of transactions, on the footing that the bid/offer would only be binding if all components were fulfilled simultaneously.

It is also clear that important conditions have to be fulfilled for such an incremental market-driven refarming process to go ahead. Because of interference problems from one service to another, guard bands are likely to be necessary. This will require bunching of similar processes in the band, to avoid waste of spectrum, and this in turn will require the spectrum regulator to have the right to "repack" the band following the two-sided auction. This is a necessary condition for the success of the version of a two-sided auction currently planned in the USA and discussed below.

To date, the closest to a two-sided spectrum auction for long-term spectrum access is the "incentive auction" proposed in the United States.[21] This envisages that an amount of 600 MHz spectrum, to be determined in the auction, will be transferred from broadcasting to mobile use. It is a key feature of this auction that the two sides of the auction

---

[19] This version of the story mirrors the notion of a multi-market equilibrium developed by the nineteenth-century French economist Léon Walras, who conceived it as arising through a price-guided process conducted by auctioneers (or dealers). But in this instance there is a clear place, which is lacking in Walras's writings, for the co-ordinating agent in the person of a real auctioneer.

[20] Here we assume that the market is effectively competitive and not dogged by such things as collusion, discussed in section 5.4 below.

[21] Some significant technical work has also been done for two-sided or double auctions for dynamic spectrum access [23].

will take place in series rather than simultaneously; that is, there will first be a "reverse auction" in which the government will purchase a relatively large number of spectrum licenses from broadcasters; then the "forward auction" of spectrum to (presumably) mobile operators will take place. The auctioneer will then compare the costs of the reverse auction with the revenues of the forward auction. If the relationship between cost and revenues is satisfactory, the surviving broadcasters will be repacked into the lower part of the relevant band, and the transfer of spectrum to mobile use will be accomplished at the relevant prices in the two auctions. If revenues are inadequate, the auctioneer reruns the reverse auction, buying fewer licenses at lower cost, and re-auctioning those forward. The process concludes when a suitable balance between costs and revenues emerges.

This ingenious and innovative process originated in the Public Safety and Spectrum Act, passed by the US Congress in 2012.[22] This Act permitted the use of auction revenues from the sale of mobile broadband spectrum for several purposes, including payments to broadcasters to release spectrum, deficit reduction, and the development of a nationwide public-safety broadband network. The latter two provisions imposed the need for a "tax wedge" between mobile auction revenues from mobile operators and compensation payments to broadcasters.

The FCC began work on implementing the Act in 2012, and released its detailed plans (Report and Order) on incentive auctions in June 2014 [24]. In the reverse auction, broadcasters have the choice of continuing to broadcast on their current band (subject to repacking), relinquishing their license altogether, or moving to or sharing another band.

A descending-clock format will be used, with broadcasters being offered in each round prices for relinquishing their license or moving to another band. The prices offered to each station for options will be adjusted downward as the rounds progress. When enough stations have dropped out so that all remaining active bidders are needed to meet the target for the amount of spectrum that is determined to be required in the next auction stage, the reverse auction for the stage will end. The price paid to each remaining bidder will be at least as high as the last price it agreed to accept.

The subsequent "forward auction" of spectrum to mobile operators will be an ascending-clock auction format. Bidders will be able to bid for generic licenses in one or more categories. Intra-round bidding will be allowed. There will be a separate clock price for each category in each geographic area, and bidders will indicate the number of licenses that they demand at the current prices. The prices generally will rise from round to round, as long as the demand for licenses exceeds their availability. Subject to certain conditions, bidders still demanding licenses when the clock prices stop rising in every license category in every area will become winners of those licenses. Winners may then indicate their preferences for frequency-specific licenses in an assignment round or a series of separate bidding rounds.

Implementation of the FCC's plan is an exacting process due to be completed in 2016, but subject to legal challenge. But it significantly extends the reach of auctions from deciding which firms get access to spectrum to deciding how the spectrum is used. It

---

[22] For a summary, see [25].

remains to be seen whether the innovation will be widely copied. Its pros and cons can be described as follows:

Pros:
- quasi-double auction approach, combining supply and demand, enables the prices emerging from the auction to determine the rate at which spectrum is switched from one use to another;
- the two sides of the auction disclose useful information about willingness to sell as well as willingness to pay.

Cons:
- the need for a target auction revenue to meet other needs makes the auction highly complex;
- its usefulness hinges on the possibility to repack broadcasters into the part of the band to be retained for that purpose; if the released spectrum were scattered throughout the band, it would not be so useful to mobile operators.

4.10 Conclusion

Spectrum auctions have been used extensively around the world for assigning many thousands of licenses covering many different uses and types of user, and to date are the most transparent and successful of the market methods applied in spectrum management. Some auctions have raised vast sums of money, while other auctions have raised very little. Competition has varied in different auctions, though the degree of competition often reflects broader market sentiments.

Since the first spectrum auction in 1989, the design of auctions has evolved to accommodate very large-scale auctions and small-scale auctions. The SAA format is often used for large-scale auctions and is particularly good where licenses may be similar, though not identical, and where there are complementarities. There is also significant use of combinatorial clock auctions.

References

[1] L. Hurwicz and S. Reiter, *Designing Economic Mechanisms*, Cambridge University Press, 2006.
[2] P. Klemperer, "Auctions: Theory and Practice" (2004), Economics Group, Nuffield College, University of Oxford, Economics Papers.
[3] P. R. Milgrom, *Putting Auction Theory to Work*, Cambridge University Press, 2004.
[4] P. Cramton, "Spectrum Auction Design" (March 2013) 42(2) *Review of Industrial Organisation* 161.

[5] P. Cramton, "How Best to Auction Oil Rights: Escaping the Resource Curse," in M. Humphreys, J. D. Sachs, and J. E. Stiglitz, eds., *Escaping the Resource Curse*, New York: Columbia University Press, 2007, 114.

[6] 2011 Report of the Comptroller and Auditor General of India for the year ended March 2010, at http://cag.gov.in/html/reports/civil/2010-11_19PA/Telecommunication%20Report.pdf.

[7] B. D. Bernheim and M. Whinston, "Menu Auctions, Resource Allocation, and Economic Influence" (1986) 101 *Quarterly Journal of Economics* 1.

[8] E. R. Kwerel and G. L. Rosston, "An Insiders' View of FCC Spectrum Auctions" (2000) 17(3) *Journal of Regulatory Economics* 253.

[9] M. Park, S.-W. Lee, and Y.-J. Choi, "Does Spectrum Auctioning Harm Consumers? Lessons from 3G Licensing" (2011) 23(1) *Information Economics and Policy* 118.

[10] See www.fcc.gov/incentiveauctions.

[11] W. Vickrey, "Counterspeculation, Auctions, and Competitive Sealed Tenders" (1961) 16(1) *Journal of Finance* 8.

[12] J. Bulow and P. Klemperer, "Auctions versus Negotiations" (1996) 86(1) *American Economic Review* 180.

[13] See www.fcc.gov/spectrum.

[14] Article 1(2) of the Radio Spectrum Decision, No. 676/2002/EC, March 7, 2002.

[15] Recital 22 of the Authorisation Directive 2002/20/EC, Brussels.

[16] See "Spectrum Auction Design in New Zealand," Ministry of Economic Development (November 2005).

[17] L. M. Ausubel, "An Efficient Ascending-Bid Auction for Multiple Objects" (2004) 94(5) *American Economic Review* 1452.

[18] J. Morgan, "Combinatorial Auctions in the Information Age: An Experimental Study," in Michael R. Baye, ed., *The Economics of the Internet and E-commerce*, Bingley: Emerald, 2002, 191.

[19] P. Cramton, Y. Shohan, and R. Steinberg, eds., *Combinatorial Auctions*, Cambridge, MA: MIT Press, 2006.

[20] G. Myers, "Spectrum Floors in the UK 4G Auction: An Innovation in Regulatory Design," at www.lse.ac.uk/researchAndExpertise/units/CARR/pdf/DPs/DP74-Geoffrey-Myers.pdf.

[21] National Audit Office, "4G Radio Spectrum Auction: Lessons Learned" (March 2014).

[22] E. Kwerel and J. Williams, "A Proposal for a Rapid Transition to Market Allocation of Spectrum," OPP Working Paper No. 38, Federal Communication Commission (November 2002).

[23] T. Alpcan, H. Boche, M. L. Honig, and H. V. Poor, eds., *Mechanisms and Games for Dynamic Spectrum Allocation*, Cambridge University Press, 2014, Chapter 15.

[24] NERA, "US 600 MHz Incentive Auction," at www.nera.com/content/dam/nera/publications/2014/PUB_600MHz_Incentive_Auc_Fwd_Rev_Auc_Rules_1014.pdf.

[25] L. Moore, "Spectrum Policy: Provisions of the 2012 Spectrum Act," Congressional Research Services (March 2014).

References

[26] P. R. Milgrom and R. J. Weber, "A Theory of Auctions and Competitive Bidding" (1982) 50(5) *Econometrica* 1089.

[27] P. Klemperer, "Auction Theory: A Guide to the Literature" (1999) 13(3) *Journal of Economic Surveys* 227.

[28] P. Cramton, "Spectrum Auctions," in M. Cave, S. Majumdar, and I. Vogelsang, eds., *Handbook of Telecommunications Economics*, Amsterdam: Elsevier Science B.V., 2002, 605.

[29] P. Cramton, E. Kwerel, G. Rosston, and A. Skrzypacz, "Using Spectrum Auctions to Enhance Competition in Wireless Services" (2011) 54(4) *Journal of Law and Economics* S167.

[30] R. B. Myerson, "Perspectives on Mechanism Design in Economic Theory" (2008) 98(3) *American Economic Review* 586.

[31] P. Milgrom, "Auctions and Bidding: A Primer" (1989) 3(3) *Journal of Economic Perspectives* 3.

[32] See http://cag.gov.in/html/reports/civil/192010-11_19PA/Telecommunication%20Report.pdf.

[33] See http://wireless.fcc.gov/auctions/data/papersAndStudies/fc970353.pdf.

[34] K. Binmore and P. Klemperer, "The Biggest Auction Ever: The Sale of the British 3G Telecom Licences" (2002) 112(478) *Economic Journal* C74.

[35] R. H. Coase, "The Federal Communications Commission" (1959) 2 *Journal of Law and Economics* 1.

[36] E. M. Noam, "Spectrum Auctions: Yesterday's Heresy, Today's Orthodoxy, Tomorrow's Anachronism. Taking the Next Step to Open Spectrum Access" (1998) 41 *Journal of Law and Economics* 765.

[37] See http://wireless.fcc.gov/services/aws/data/awsbandplan.pdf.

[38] See http://wireless.fcc.gov/auctions/data/maps/reag.pdf.

[39] See http://media.ofcom.org.uk/news/192013/winners-of-the-4g-mobile-auction.

[40] See www.acma.gov.au/Industry/Spectrum/Digital-Dividend-700MHz-and-25Gz-Auction/Reallocation/digital-dividend-auction-results.

[41] See https://apps.fcc.gov/edocs_public/attachmatch/FCC-14-50A1.pdf.

5 Other aspects of spectrum auction design

5.1 Introduction

In Chapter 4 we examined the economics of spectrum auctions, considered why auctions have become so prevalent, and looked at the good and bad features of a number of auction variants. In this chapter we look at further processes associated with auction design and implementation.

We begin by examining the logistics of an auction. This includes the invitation and pre-qualification stages, and the subsequent grant of a license. The next issue concerns the design of the lots to be used in the spectrum award; these are based on bidders' needs, the technology likely to be deployed, and the type of auction being used. We then turn to ways of encouraging a competitive auction and the interaction between the design of that auction and competition law. We end by considering the interaction between auctions and downstream competition.

5.2 Auction logistics

An auction is one part of a four-stage assignment process:

1. an invitation stage,
2. a pre-qualification stage,
3. an auction stage, and
4. a grant stage.

These stages follow a sequence, from promoting awareness of the auction among potential bidders (marketing) to the eventual assignment of licenses to successful bidders (the grant stage). This section describes in greater detail each of the stages involved.

5.2.1 The invitation stage

The invitation stage usually comprises the publication of an information memorandum, containing details about all subsequent stages of the auction, application forms and pre-qualification requirements. This stage is an important part of the overall process as it includes the period in which to market the auction. If revenue is an important criterion, then participation will be important.

5.2.2 Pre-qualification stage

The pre-qualification stage in an auction provides an opportunity for the spectrum agency to screen out "inappropriate" bidders and to learn more about likely demand. This is achieved by setting out criteria that must be satisfied by all prospective "appropriate" bidders before they are accepted as participants. In effect a pre-qualification stage is a hurdle to participation. No comparison is made across bidders, distinguishing it from a beauty contest or comparative selection process.

The criteria used should ideally be objective and transparent, and not impose a significant burden on prospective bidders. After all, one of the main purposes of an auction is to delegate assignment to competitive forces rather than to bureaucracy. For example, in the Nigerian GSM auction the pre-qualification criteria comprised a few straightforward questions, selected principally to

- deter those engaged in money laundering (a major concern in Nigeria) or other illegal activities from participating,
- ensure that bidders satisfied ownership rules designed to stop bidder collusion, and
- ensure that bidders had some experience of delivering telephony services.

In other jurisdictions the focus is on requiring bidders to offer proof that they have the funds to take part.

As part of pre-qualification it is usual in spectrum auctions to require would-be bidders to deposit a sum of money with the auctioneer. By requesting up-front deposits, financial penalties for breach of auction rules can be made more credible. Deposits are also used by auctioneers as a part payment towards the objects won by successful bidders, and to this end can reduce the risk and offset the costs associated with default.

Deposits, and more generally the payment terms for an object won in an auction, affect the degree of speculative activity that may occur in an auction. Speculation tends to arise in situations where information is imperfect, and particularly where some bidders are better informed than others – the case of asymmetric information. It can also occur where the rules of an auction are biased towards favoring participation by certain types of bidder, giving rise to asymmetry among participants.

Speculative bidders seek to win objects with the goal of selling them on subsequently at higher prices. As the market price of an object won in an auction is not typically known with certainty prior to an auction, speculators may default on paying for an object won in an auction if it becomes apparent that the market price is below the auction price. The level chosen for the deposit can help reduce speculative activity, and is therefore an important instrument available to auction designers.

In practice, an auctioneer operates in a far more complex and uncertain setting than is suggested in this example. Nevertheless, it usually remains the case that a higher deposit provides a greater deterrent to speculative bidding.

As a further deterrent to speculative activity, it is common practice to require the successful bidder to pay the license fee in full within a given period following the auction closure. If a successful bidder defaults on payment, various measures can be taken, such as banning the non-payer from any future license assignments for a period of years and retaining the bidder's deposit.

If the risk of default is a concern in a spectrum auction, other measures can reduce it further. One common instrument that has been used in a number of spectrum auctions is the requirement that bidders submit bank guarantees before, and even during, an auction. This requirement places the risk-assessment burden onto banks. As banks are better informed about the risk characteristics of bidders, the probability of default is likely to be lower in the presence of bank guarantees. The insistence on bank guarantees is sometimes welcomed by "serious" bidders, as it means that the auction price is more likely to reflect fundamentals and will not be inflated by insincere bidding.

In advanced economies with mature capital markets, shareholder governance will usually ensure sincere bidding in high-stakes auctions. In economies with less sophisticated capital markets, monitoring of management by shareholders is likely to be less effective at preventing insincere bidding. In such circumstances there may be good reasons to insist upon bank guarantees.

5.2.3 Auction stage: designing the rules

Auction rules are often very detailed and cover activities by bidders before, during and immediately after the auction. Rules are required to prevent collusive behavior that could undermine efficiency, and to provide detailed guidance about what is and what is not permitted during an auction. Detailed and precise rules are required in order to give bidders as much certainty as possible, encouraging them to concentrate on valuing licenses rather than engaging in excessive strategizing against each other. Spectrum auction rules are typically drafted by economists and lawyers, working together with the auctioneer.

Auction designers often specify rules for almost all eventualities, even for circumstances that may seem highly unlikely. Detailed rules are required because should an unlikely event occur to which no rule applies, the process falls into disarray, as rules have to be made up on the spot. Furthermore, it is vital that bidders know precisely what the consequences will be for any action they take. Otherwise the uncertainties can cause them to misbehave either intentionally or unintentionally. For example, if there were some action a bidder could take of which the consequences were not pre-specified, they could attempt to game the system by doing that and throwing the auction process into the courts to be resolved. This has been attempted numerous times in FCC auctions (though with little success).

In many SMRA spectrum auctions, bidders are given an opportunity to pause and reflect on their strategy. This is usually permitted in two ways: via "waivers" and through the calling of a "recess." The action of a waiver is where a bidder is allowed to abstain from making a bid in a round. A waiver is intended to give a bid team pause for thought and possibly time to communicate with financiers and other interested parties.

Whether or not to allow waivers in an auction poses a challenge for the auction designer. Bidders expect and should be granted a reasonable amount of time to discuss their valuation, particularly as bidders in high-stakes spectrum auctions are typically consortia whose members may have different views and means of access to capital. However, allowing bidders time to compose bids by granting waivers means that there is a chance they could use waivers to mislead other bidders, or to take advantage of information revealed in an auction by other bidders. In the UK 3G auction, bidders were permitted a maximum of three waivers and one recess day.

Bid increments need to be determined by the auctioneer and can be chosen to accelerate the pace of an auction if bidding is too slow.

5.2.4   The grant stage

The successful bidders in an auction progress to the grant stage, when licenses are issued subject to receipt of monies owed to the auctioneer.

5.3   Lot design

5.3.1   Overview

In auctions, lot design helps to ensure that there is an attractive set of options for bidders. This makes the auction more efficient, irrespective of the design of the auction itself. Lot design presents a challenge for the auction managers, who can draw on their own knowledge and international outcomes of auctions for similar spectrum. However, the success and efficiency of an auction will turn on designing lots that are attractive to the firms which are likely to bid. This means holding discussions with potential bidders both in private and, possibly, in public if this does not require potential bidders to disclose their private information to other potential bidders.

For example, if each of the lots in a particular auction were 2 × 5 MHz with a total of 2 × 30 MHz available and a spectrum cap (see section 5.5.3 below) of 2 × 20 MHz, then information regarding the pre-qualification deposits paid by the bidders could reveal bidding intentions in a way which could be either beneficial or harmful to the auction's efficiency, depending on the auction design. Whatever disclosure is decided upon, it is crucial that the spectrum manager maintain confidentiality and enforce confidentiality obligations on others; see section 5.4.4 below.

5.3.2   The myth of technology-neutrality

In some countries, such as Australia and New Zealand, spectrum is auctioned without a requirement that it be used for the delivery of a particular form of service or that it be used for a specific technology. In most European countries, spectrum is auctioned with a specific use mandated and often with rollout obligations associated with the specified technology.

Whether the technology is specified or otherwise, it is essential that the lot design reflect likely applications. For example, if the spectrum to be auctioned is paired, then there is no utility in having separate lots for the separate blocks that form the pair.

Similarly, there is no benefit in designing lots which have spectrum pairings that are different from the approach taken by neighboring states.

The minimum size of a lot (whether paired or unpaired) is likely to reflect the smallest amount of spectrum which is likely to be independently valuable, taking into account the likely technology that will be deployed. Although technologies which allow the aggregation of spectrum in multiple bands such as LTE-A are emerging, designing lots on the assumption that customer equipment will emerge over time could be problematic. For technologies that were mainstream in 2015, this means that the most likely lot size will be a multiple of 2×5 MHz or 5 MHz.

5.3.3 Working with potential bidders

As an auction is simply a mechanism by which buyers and sellers meet in a marketplace, it is important that the auction manager, the seller, have a clear understanding of the likely needs of the buyers. Necessarily, this means working together with the potential bidders in an auction from an early stage; that is, well before the information memorandum concerning the auction is released.

This interaction between auction manager and bidders ranges from marketing the lots that are on offer through to ensuring that the auction lots are attractive to bidders.

It is critical that auction managers in their interactions with the potential bidders do not allow potential bidders to exchange private-value information in a way that would adversely affect the efficiency of the auction. The worst outcome would be for an auctioneer to facilitate the type of information exchange that is described in section 5.4.2 below.

5.3.4 Lot design and auction type

To some extent, there is a connection between lot design and the type of auction. Consider the sale of 2×30 MHz of spectrum where the lot size is 2×5 MHz. If the lots are defined in terms of lower and upper bounds, then there may be (and is likely to be) an additional value for having contiguous lots. This value could be expressed by having a combinatorial auction or by having a second phase to a clock auction (or SMRA) where the value of contiguity can be expressed.

In each case, the simultaneous auction of the lots is important in order to improve the chances that there will be an efficient outcome without a "winner's curse" and without the opportunity for another bidder to drive up a price without intending to win the lot.

5.4 Ensuring a competitive auction

5.4.1 How competition law affects auctions

Competition, or antitrust, law is primarily concerned with increasing consumer welfare by promoting competition. This promotion of competition occurs through orderly markets that are not distorted by anticompetitive conduct. In order to minimize such distortions, competition law has three pillars:

1. prohibitions on anticompetitive co-ordinated conduct,[1]
2. prohibition on the unilateral abuse of market power, and
3. merger control.

Cartels or co-ordination occur when two or more firms enter into agreements or act in parallel in order to share markets or fix either the price or non-price terms of either supply or acquisition. The unilateral abuse of market power is conducted by a single firm which relies on the abuser having sufficient market power for the abuse to be effective. Merger control is a regulatory mechanism to prevent the acquisition of an asset (usually, but not always, ownership of an enterprise) from resulting in a substantial lessening of competition in any market. In designing an auction for the assignment of spectrum, there is a need to address each of these issues.

There is a question as to the extent to which design of the auction process needs to operate separately from the operation of general competition law. The answer to this question depends on the jurisdiction. For example, when the first 3G spectrum auctions took place in Hong Kong, there was no general competition law and a significant part of the auction process design was to ensure that there was no co-ordination among bidders and potential bidders. Where there is a competition law which prohibits co-ordinating or cartel behavior, it would not in normal circumstances be disapplied in an auction. Thus in Australia for the 2013 auctions of 700 MHz and 2.5 GHz spectrum, the "Auction Guide" [1] noted, "the rules are intended to guard against anti-competitive behaviour in the auction and to complement the prohibition on cartel conduct ... [provided by law]."

5.4.2 Co-ordinated conduct

The risk of co-ordinated conduct occurring in spectrum auctions arises in part because the amounts of money involved are potentially extraordinarily high. The Federal Communications Commission auction for spectrum in the 1.8 GHz band that finished in the US in 2015 raised more than $45 billion. Any bidder in such a high-stakes game would, acting rationally, seek to minimize its expenditure by any legal means.

This risk is heightened when the auction design facilitates co-ordination at any point in the process. For example, if there are three incumbent mobile operators and there is an auction for 2 × 30 MHz of spectrum with lots of 2 × 5 MHz, then there is a significant likelihood that the outcome will be that each operator acquires 2 × 10 MHz. However, there is some risk that this outcome could be as a result of co-ordination. It is therefore important that the auction process design should either

---

[1] A particular example of such conduct is operation of a bid-rigging cartel, which in many jurisdictions invites criminal penalties. On bid-rigging in general, see [2].

minimize this risk or create a simple way for co-ordinated bidders to defect from the prior agreement.

That is, the auction manager should be aware of the types of strategy used by competition regulators to encourage cartel defection. This is because the evidence suggests that defection is a common cause of cartel breakdown, even though other factors, such as entry, exogenous shocks, and bargaining problems are also often encountered [3].

Auction design can facilitate co-ordinated conduct in the auction itself. Cramton and Schwartz describe the bid signaling in FCC spectrum auctions. They found that simultaneous open bidding allowed bidders to send messages to their rivals "telling them on which licenses to bid and which to avoid. This 'code bidding' occurs when one bidder tags the last few digits of its bid with the market number of a related license." They also found that "only a small fraction of the bidders commonly used retaliating bids and code bids. These bidders won more than 40% of the spectrum for sale and paid significantly less for their overall winnings" [4, pp. 4, 14]. It is harder to argue that there has been wrongdoing or a breach of the auction rules if the auction design itself facilitates the conduct about which there is a complaint.

One specific aspect of the auction design that will assist in minimizing the potential for co-ordinated conduct is for the auction manager to have a clear understanding of the potential bidders and their commercial relationships with each other. This is often achieved by requiring that any bidders with significant common ownership or common directors should be regarded as a bidding group – that is, they form a single bidding unit with an expectation that any spectrum trading within the group will occur after the auction.

This is achieved by setting ownership disclosure obligations as part of the pre-qualification stage and imposing an ongoing obligation to disclose changes in the composition of bidding groups from time to time as soon as is practicable after the change occurs.

5.4.3 Setting a reserve price

A reserve price in an auction is a price below which the item will be withdrawn from sale. An argument can be made that there is no need to set a reserve price in an auction where the objective is to achieve the economically most efficient outcome. The auction process will find the correct price and the setting of a reserve may adversely impact the price-discovery process. However, the setting of a starting price, usually the same as the reserve price, reduces the time required for an auction and forms part of the price-discovery process. It is also often convenient for the auction manager to have a reserve price in order to set the requirements for deposit bank guarantees and other funding processes.

If there is a risk of collusion, a reserve sets a minimum price which the colluders must pay, and places a limit on the extent to which they can appropriate for themselves the scarcity value of the spectrum, rather than leave it with the government. Moreover, our discussion above at section 4.7.3 of 3G auctions in Europe showed how, as the auctions successively unfolded, the number of participants fell until they equalled the number of licenses available. In such a case, absent a reserve price, the auction might elicit purely token payments.

There are potential problems in setting the reserve price at a level which is too high. One effect is to reduce the number of bidders. For example, Vodafone in Australia withdrew from the 2013 auction for 700 MHz spectrum once the reserve price had been set. One problem here is that governments use auction reserve prices in forming their budgetary projections of sales revenues, and hence indirectly their public-spending plans; this may impose upward pressure on them.

Serious problems can arise if reserve prices are set at a level which permits some licenses to be sold, while others are not. Firms which have paid the reserve price naturally object very strongly if the unsold licenses are then offered at lower prices. In Europe there may also be issues concerning State Aid rules. As a result, access to valuable spectrum can be lost for many years. This outcome was observed with respect to the four 3G licenses offered for sale in France in 2001. The initial price was set too high, and as a result the fourth and final license was only granted in 2010 [5].

The auction manager is likely to need to make an assessment of a suitable starting price and will use discussions with potential bidders (which will usually point to a lower starting price) and international benchmarks of actual outcomes (expressed as $/MHz/population) in countries with comparable gross domestic products. Section 7.6 outlines how such calculations can be made in practice.

5.4.4 Management

At a pre-auction stage the auction manager will normally issue some form of guide to the auction being conducted. This is sometimes referred to as an "information memorandum" or an "auction guide." This document often incorporates, either explicitly or by reference, the associated legislation, subordinate legislation, and other regulations under which the auction is being held.

The information memorandum is a document that aims to provide marketing information concerning the spectrum to be auctioned, as well as the process for the auction itself. The document will typically also include a disclaimer to indicate

a. that the reader should take advice on the material presented,
b. that the auction may not take place at all,
c. that the pre-qualification steps set out in the information memorandum must be completed in order for a potential bidder to bid, and
d. that any timetable is indicative and not binding on the auction manager.

The other typical elements of an information memorandum are set out in Table 5-1.

Table 5-1. Typical elements in an auction information memorandum

Element	Contents
Overview	The overview acts as a form of summary of the auction, including providing information on the auction format, any sequencing of auction phases, a summary of the regulatory obligations and opportunities in the country, the spectrum offered in the auction, information on future spectrum release, and the auction timetable
Country-specific information	This section will typically set out the environment into which a potential bidder will be offering services. It is likely to set out the spectrum held by others (or a reference to where that information may be found), as well as information on the vibrancy of competition in the sector. This section is particularly important if the auction is intended to encourage new entrants
License terms	The terms of the licenses being offered include the boundaries expressed as regards frequency, geography, and time (that is, the duration of the license). This may include specific regulatory obligations or opportunities such as coverage obligations or roaming rights. The payment terms outside the auction payments (for example, contribution to a universal service fund) may also be included
Regulatory obligations	This section sets out more general regulatory obligations on matters such as number portability and infrastructure sharing
Pre-qualification	This section sets out the requirements for prospective bidders to qualify to take part in the auction
The auction	This section is the heart of the description of the auction process and will usually restate the timetable and the spectrum being sold before describing both the conduct of the auction and the auction procedure
Confidentiality and other terms	This section sets out the confidentiality regime that will bind prospective bidders, including from when the confidentiality obligations apply. It will also usually provide information on matters such as: • disqualification, • ownership disclosure requirements, • changes in composition of bidders, and • collusion

5.4.5 Enforcement

The auction manager will need to be able to enforce the rules of the auction. As a result, the auction manager requires a range of punishment options for noncompliance. These will normally range from forfeiture of some or all of the deposit through to disqualification (including loss of deposit). The auction manager is likely to need an explicit right to litigate for at least breach of contract in the event that the rule breaking was egregious.

5.5 Auctions and downstream competition

5.5.1 Introduction

If the goal of auction design were to maximize revenues, this would probably best be achieved by auctioning all of the spectrum capable of delivering a particular service to a single provider, rather as medieval kings maximized revenue by auctioning salt monopolies. The winner would then be the most efficient monopolist, but a monopolist nonetheless.

Accordingly auction designers must have an eye to the effects of an auction, or of a sequence of auctions, on competition in downstream markets. When auctions began to be used to assign spectrum for mobile communications, the auction was designed to sell a specified number of licenses, each with a preassigned number of megahertz, on a "one-operator, one-license" basis. The number of licenses was controlled, and typically increased over time to inject more players into the marketplace. As several frequencies came to be available to provide mobile voice and data (so that operators built up portfolios of spectrum holdings), and as spectrum came to be sold in smaller units via clock auctions (thus making the spectrum market more competitive at the margin), regulators had to address the implications for competition of auction design more explicitly.

The nature of the problem was well expressed by the US Department of Justice, in its comments on the operation of the proposed US "incentive auction," discussed in Section 4.9. It noted that large or dominant incumbents in a spectrum-using sector such as mobile communications ascribe private value to the spectrum not only on account of the revenues such assets will bring in, but also because the incumbents' acquisition of spectrum rights will prevent rivals from eroding those incumbents' existing profits: "The latter might be called 'foreclosure value' as distinct from 'use value' … The 'foreclosure value' … represents the private value of foreclosing competition by, for instance, forestalling entry or expansion that threatens to inject additional competition into the market" [6].[2]

The scope in a spectrum auction for foreclosure of a downstream market depends crucially upon which frequencies are good substitutes for one another in the provision of a particular service, such as mobile communications. If many frequencies can provide

---

[2] Foreclosure denotes action taken by a firm expressly to prevent a competitor from entering a market, or to weaken it in the market. Ofcom refers to this aspect of spectrum value as "strategic investment value" [6].

the service at the same quality level, and if it is possible to substitute a non-spectrum input for spectrum in addition,[3] then the chances of a successful foreclosure are diminished. Equally, if there are a large number of contiguous frequencies exhibiting gradually changing properties, there can be a "continuous chain of substitution" across the whole lot.

The question is thus one often encountered in competition analysis: are different frequencies to be regarded as close enough substitutes to fall in the same market? Conceptually this question is best answered by asking whether a hypothetical monopolist, the sole firm controlling access to, say, mobile spectrum under 1 GHz, would be able to raise the price of such spectrum by, say, 10% above the competitive level. In practice the answer to the question tends to revolve around the capability of different frequencies to provide coverage and capacity for mobile services of a specified quality (for example, in terms of building penetration), and the observed behavior of parities in spectrum auctions.

The frequency allocation table in many countries confines mobile spectrum to a small but growing number of principal frequencies. In many European countries these are at 800/900 MHz, 1800 MHz, 2.1 GHz, and 2.6 GHz. The significant gap between the sub-1 GHz spectrum and the other bands has led many spectrum regulators to adopt a different treatment of these two categories, without necessarily having examined the matter in detail. But the distinction is bolstered by the observation that the price realized at auction by sub-1 GHz spectrum is several times greater than that realized at 1800 MHz.

Ofcom's reasoning for distinguishing sub-1 GHz spectrum from the rest, for mobile communications purposes, is as follows:

Particular importance of sub-1 GHz spectrum
5.40 Sub-1 GHz spectrum gives advantages over higher frequencies in terms of coverage. It allows a significantly greater geographical area to be served than higher frequency bands would, for the same number of sites (because signals travel further at lower frequencies). It also tends to provide substantially better signal quality and higher download speeds (throughput) within buildings than higher frequencies since lower frequency signals are better at penetrating solid objects.

5.41 These advantages could mean that national wholesalers with a large amount of sub-1 GHz spectrum would have an unmatchable competitive advantage over those without any sub-1 GHz spectrum. By an unmatchable competitive advantage we mean that the national wholesalers without sub-1 GHz spectrum suffer a material competitive disadvantage because they are unable to develop their networks to offer services sufficiently similar to national wholesalers with sub-1 GHz spectrum. This would depend partly on technical differences between wholesalers with different spectrum portfolios and partly on how sensitive consumers are to any such technical differences, such as the quality of deep indoor coverage.

5.42 ... Our preliminary conclusion is that national wholesalers with a large amount of sub-1 GHz spectrum may have an unmatchable technical advantage compared to national wholesalers without any sub-1 GHz spectrum.[4]

---

[3] This can be accomplished to some degree by designing a network with more base stations, operating at lower power levels. This allows spectrum to be reused at lesser distances, thus reducing the spectrum holding required to serve a given area.

[4] Ofcom, "Consultation on Assessment of Future Mobile Competition and Proposals for the Award of 800 MHz and 2.6 GHz Spectrum and Related Issues," March 22, 2011. In its 2013 auction design, which incorporated a spectrum floor for a fourth bidder, Ofcom effectively guaranteed that bidder a choice among

5.5.2 How competition can be encouraged

A spectrum regulator concerned about market power in the provision of spectrum has access to a number of possible instruments to deal with it [8, 9]:

1. **Competition law.** Reliance can be placed on competition law to prevent the acquisition of spectrum in an anticompetitive manner. The problem here is that a transaction in which a spectrum-using company buys spectrum from another such company may not be captured by competition law.[5]
2. **A "use-it-or-lose-it" rule.** Purchases of unnecessary spectrum can be discouraged by adopting such a rule, which requires a firm to hand back spectrum assigned to it but not put to use within a specified period. This is intended to discourage hoarding of spectrum. The problems here are that the rule may discourage the acquisition of spectrum ahead of need even where acquisition is rational, and that a rule requiring use is difficult to monitor and easy to circumvent.
3. **Vetting spectrum trades.** Trades can be vetted before they occur. This is discussed in Chapter 6 below. But it only covers secondary or resale transactions, and does not cover primary or auctioned awards.
4. **Set-asides.** Some spectrum can be made available for acquisition only by bidders satisfying particular conditions – for example, being new entrants into the relevant marketplace. This encourages participation by operators in the favored class, but it may have the effect of assigning spectrum to less efficient operators. This method has been used both in North America and in some European countries, with mixed results.
5. **Bidding credits.** Under this arrangement, bidders of a favored type are treated as if they had bid more money for the spectrum than in fact they have to pay. For example, new entrants may get a 25% bidding credit. In this case, a new entrant submitting a bid which would oblige it, if successful, to pay $80 would be treated for the purpose of assigning the spectrum "as if" it had bid $100. Care must be taken to prevent a favored bidder from simply realizing a capital gain by immediately trading the spectrum to an unfavored buyer.

However, the most commonly used method is to employ spectrum caps. Either these place a restriction on the quantity of spectrum which any operator can hold in any (or in all) frequencies, however and whenever acquired, or the limit can apply to what is acquired in the course of a particular spectrum award. The first type of constraint may be a "soft cap"; that is, a prospective breach is not automatically prohibited, but may

---

several spectrum bands. From those available, the fourth bidder elected to acquire 800 MHz spectrum; see Box 5-1 below.

[5] This varies from jurisdiction to jurisdiction. Thus acquisition of spectrum is treated as a merger situation in Australia but not in the law of the European Union. However, the acquisition of a spectrum-using enterprise, such as a rival mobile operator, would normally be so covered, and in fact such acquisitions in the European Union have sometimes only been allowed to go ahead if the combined entity agrees to divest itself of some spectrum. See section 5.5.4 below.

trigger an investigation. But an auction-specific cap is likely to be nonnegotiable: it is a tied-down element of the auction rules. And because they are a feature of the rules of a specific auction, the spectrum regulator has some discretion over what rules to impose.

5.5.3 Spectrum caps

Spectrum caps in auctions are a double-edged sword: they have the potential to prevent anticompetitive conduct by large firms, but, if they are imposed unwisely or too harshly, they can penalize successful operators which have won a large number of customers and hence have a need for substantial amounts of spectrum.

If caps in an auction do indeed have the effect of redirecting spectrum away from operators to whom it is valuable in favor of less efficient firms, it is also possible that the resulting loss of efficiency will be reflected in reduced revenues. However, against this it can be argued that, absent caps, smaller operators will assume that they have no realistic chance of acquiring spectrum and not participate in the auction, reducing demand for the lots available. An educated guess is required to decide which of these two effects will predominate.

In the United States a heated debate has taken place over the use of caps in auctions [10]. But in recent auctions in Europe, especially those occurring since 2010 and involving the sale of 800 MHz spectrum released as part of the switch-over from analogue to digital terrestrial television (DTT), spectrum caps have been used almost universally, in application to the sub-1 GHz spectrum made available. In some cases a separate cap was applied to spectrum above 1 GHz; in others a limit was imposed on an operator's total acquisition of spectrum.[6] See Table 5-2.

The UK auction included in Table 5-2, and discussed in Section 4.8 above, was also remarkable in that it contained not only a spectrum cap, but also a special arrangement known as a spectrum floor, which was designed to prevent any of the four mobile operators from being placed at a major competitive disadvantage as a result not having access to sufficient spectrum of suitable quality to be "a credible national wholesaler of mobile services." This is described in Box 5-1.

5.5.4 Consolidation, mergers, and input sharing

As was discussed above, one of the fundamental pillars of competition law is merger control. This raises an issue concerning how auction caps should be treated in any merger that occurs after the auction or in any post-auction spectrum trade. After all, if caps imposed at an auction are expressly designed to provide a pro-competitive outcome, this can be taken to imply that the subsequent amalgamation of two operators' holdings would run the risk of lessening competition in the downstream market. However, post-auction mergers and trades are not restricted by caps which only apply in the auction phase.

---

[6] In the Netherlands, some spectrum was also "set aside," with unsatisfactory results.

Table 5-2. Summary of caps in European auctions and outcomes

Country	Number of national MNOs	Aggregation limit in auction	Sub-1 GHz limit	Outcome
Germany	4	2 × 20 MHz (2 × 22.4 MHz Vodafone and T-Mobile) below 1 GHz	Yes	2 × 10 MHz (O2); 2 × 10 MHz (T-Mobile); 2 × 10 MHz (Vodafone)
Ireland	4	2 × 20 MHz	Yes	2 × 10 MHz (Meteor); 2 × 10 MHz (Vodafone); 2 × 10 MHz (O2)
Switzerland	3	2 × 25 MHz on combined 800 MHz and 900 MHz	Yes	2 × 10 MHz (Orange); 2 × 10 MHz (Sunrise); 2 × 10 MHz (Swisscom)
Sweden	4	2 × 10 MHz	Yes	2 × 10 MHz (3); 2 × 10 MHz (Telenor); 2 × 10 MHz (Sulab)
Spain	4	2 × 20 MHz	Yes	2 × 5 MHz (Telefónica); 2 × 5 MHz (Vodafone); 2 × 20 MHz (Orange)
Portugal	3	2 × 10 MHz	Yes	2 × 10 MHz (Optimus); 2 × 10 MHz (TMN); 2 × 10 MHz (Vodafone)
Slovakia	3	2 × 10 MHz	Yes	Auction due 2H 2013
Iceland	6	2 × 20 MHz	Yes	Winners were Nova, Vodafone, Simmin, and 365 (amounts unknown)
Czech Republic	3	2 × 15 MHz with roaming provision	Yes	Bidding halted because prices considered too high
UK	4	2 × 27 MHz below 1 GHz plus floor and coverage obligation	Yes	2 × 5 MHz (EE); 2 × 5 MHz (3); 2 × 10 MHz (Vodafone); 2 × 10 MHz (Telefónica)
Italy	4	2 × 25 MHz below 1 GHz	Yes	2 × 10 MHz (TIM); 2 × 10 MHz (Vodafone); 2 × 10 MHz (Wind)
Norway	6	2 × 10 MHz	Yes	2 × 10 MHz (TeliaSonera); 2 × 10 MHz (Telenor); 2 × 10 MHz (Telco Data)
France	4	2 × 15 MHz with roaming provision	Yes[1]	2 × 10 MHz (Bouygues); 2 × 10 MHz (Orange); 2 × 10 MHz (SFR)
Netherlands	5	N/A (blocks set aside for new entrants)	Yes (via set-asides)	2 × 10 MHz (Tele2)
Denmark	6	2 × 20 MHz	Yes	2 × 20 MHz (TDC); 2 × 10 MHz (TT)

[1] France also imposed roaming obligations on low-band incumbents.
Source: [11].

> **Box 5-1.** Spectrum floors in the UK 4G auction
>
> In the UK auction for 800 MHz 4G spectrum, and other bands, the UK spectrum regulator Ofcom set itself the goal not only of avoiding a highly asymmetric distribution of spectrum after the auction with spectrum caps, but also of ensuring the flexible reservation of spectrum for new entrants or the smallest incumbent, to promote downstream mobile competition between at least four national mobile competitors.
>
> The latter goal was accomplished by giving either the smallest incumbent or a new national entrant the opportunity to hold, following the auction, a portfolio of spectrum which would allow it to play its part in competition in the mobile marketplace.
>
> Allowing an entrant to acquire pre-specified spectrum at a favorable price has been tried in several countries by means of set-aside (see above). But it was the regulator which determined what spectrum was made available. The innovation in the use of spectrum floors in the UK auction was that it embodied two dimensions of flexibility: different portfolios of spectrum could be reserved for different players, e.g. depending on their pre-auction spectrum holdings; and the choice of spectrum to be reserved from a range of portfolios, each of which is sufficient to promote competition, is decided through the auction as the floor that minimizes the loss in bid value from reservation. This was accomplished by adding extra rules and procedures onto an otherwise fairly standard combinatorial clock auction.
>
> In the event, the smallest incumbent operator, H3G, succeeded in acquiring 2 × 5 MHz of sub-1 GHz spectrum at Ofcom's low reserve price. After the auction, the board of H3G was said to be incredulous that the regulator's scheme, complicated as it was, had delivered the intended outcome.
>
> Source: [12]

In practice, there is a significant degree of consolidation occurring in the mobile-telecommunications sector on a global basis and this has proceeded with approval from relevant competition regulators (see, for example, [13]). In many countries this means a reduction in the number of mobile network operators from four to three. Table 5-3 shows European mobile mergers recently cleared by the European Commission. Divestment of spectrum holdings has been required in some cases. There is a marked divergence of views about the desirability of such consolidation, even among European regulators.

A further important trend in the mobile sector is towards input sharing by operators. This can take the form either of two or more operators sharing passive inputs, such as towers or antennas, or of operators combining their radio-access networks and even the spectrum holdings upon which they rely. European competition regulators initially expressed concern that more thoroughgoing versions of network sharing might create conditions for retail market co-ordination. More recently, however, a more permissive approach has been adopted [15].

Table 5-3. Consolidation of mobile operators in Europe: cleared mergers

Year	Country	Parties	Press release
2006	Austria	T-Mobile and tele.ring	IP/06/535
2007	Netherlands	T-Mobile and Orange	IP/07/1238
2010	UK	T-Mobile and Orange	IP/10/208
2012	Austria	Hutchison 3G and Orange	IP/12/1361
2014	Ireland	Hutchison 3G and Telefónica Ireland (O2 Ireland)	IP/14/607
2014	Germany	E-Plus and Telefónica Germany (O2 Germany)	IP/14/771

Source: [14].

5.5.5 Bringing the elements together

Some spectrum-using markets, especially mobile communications, appear to be prone to domination by small numbers of firms. Indeed, the structure of the mobile sector has been powerfully affected by spectrum licensing policy. For several decades this operated to promote infrastructure competition, but a trend towards consolidation and infrastructure sharing is now observable.

In these circumstances, auctions can become an unwitting instrument for exacerbating market failure, as firms use them to foreclose competition. There are therefore grounds for using spectrum rules, especially caps and floors in auctions, to limit these possibilities. Unfortunately, such interventions can lead to "regulatory failure," and spectrum regulators must take care to avoid this bad outcome.

5.6 Can demand for unlicensed spectrum be accommodated in a spectrum auction?

Bidders in spectrum auctions are typically seeking exclusive access to a band, which guarantees them interference-free access to spectrum and denies access to others who may be competing with them in downstream markets. Unlicensed spectrum, by contrast, is available to all potential users who satisfy certain regulatory conditions – for example relating to limitations on transmission power. Protecting users from interference has traditionally been accomplished by requiring low power and a geographically limited range. Increasingly it will be accomplished via databases which notify users whether there is spare capacity on the band. Alternatively, a firm might choose to bid for spectrum to create a so-called "private commons."

This distinction places individual licensees in a position where they can monetize their spectrum use, while unlicensed spectrum use cannot be so monetized. So far this has prevented competition for spectrum access between the two types of user, and left the allocation decision between them a matter for "command-and-control" regulation. This situation is likely to persist as discussion proceeds about the possible assignment of additional spectrum at 5 GHz and elsewhere to unlicensed use.

But while access to unlicensed spectrum does not have a price, it clearly has a value. Is there a way we can derive this value? We could interrogate potential users about the value they place on access to unlicensed spectrum, but they might have an incentive not to be truthful if they thought either that giving an exaggerated valuation would make it more likely that the unlicensed access would be made available, or that giving a lowered valuation would reduce the chance of being asked to make a contribution to the cost.

This problem of truthful disclosure of private valuations of public goods is a very general one.[7] As we have seen in Chapter 4, a second-price sealed-bid auction exhibits the so-called Vickrey property of eliciting truthful valuations, and it can be generalized to the case of public goods.[8] However, there are both theoretical and practical problems in implementing it. Giving people an incentive to give truthful responses may require making side payments to certain of the parties involved which may make the process very expensive.

More fundamentally it requires the identification and interrogation of a large group of possibly widely dispersed organizations which may have an imperfect understanding of what it is that they are valuing, and which may experience difficulty in projecting how their valuations are likely to change in the future [16, p. 25]. Unfortunately, the problem of eliciting truthful disclosure of valuations may make inoperable a mechanism which otherwise might work satisfactorily.

However, this does not exclude the incorporation in an auction of an agent, or "public trustee," who might bid on behalf of unlicensed users, relying on estimates of the value of such access. A number of studies have supplied such estimates [17,18]. Typically they proceed by identifying individual services which can be supplied by unlicensed spectrum, such as residential Wi-Fi, Wi-Fi cellular off-loading, RFID,[9] and so on, establishing current levels of use and value, and projecting them forward over the next 10–20 years. An agent with this knowledge could then be authorized by a public authority to bid up to this calculated amount for the spectrum, in competition with other categories of user.

However, the damage to the public finances of assigning spectrum for this purpose immediately becomes crystal clear: whereas a firm acquiring an exclusive license or using the spectrum as a private commons would be expected to pay the finance ministry for it, there would be no obvious source of payment for use of the unlicensed spectrum.

A further complication is that to get full advantage of unlicensed spectrum some form of international co-operation is required. This may prove elusive if the quasi-market instrument of introducing a "public trustee" bidder is employed.

---

[7] A public good is a good, such as national defense, which has the two characteristics (i) that consumers do not compete with one another when using it and (ii) that they cannot be excluded from using it where it is made available. Respondents asked how they value a public good may give inquirers either an artificially high valuation, to make sure it is produced in large quantities, or an artificially low one, if they are concerned that if they gave a high valuation they will be assessed for a higher share of the costs.

[8] For a description of how to apply such a mechanism in the context of spectrum management, see [19].

[9] Radio frequency identification (RFID) is the wireless use of electromagnetic fields to transfer data, for the purposes of automatically identifying and tracking tags attached to objects.

5.7 Conclusion

This chapter has examined aspects of spectrum auction process design in the context of achieving both a competitive auction process and a competitive environment after the auction has been completed and the associated licenses have been issued. In the next chapter we examine how competition and the operation of market forces can be sustained beyond the primary auction process.

References

[1] See www.acma.gov.au/~/media/Spectrum%20Outlook%20and%20Review/Information/pdf/Auctionguide%20pdf.pdf.

[2] R. C. Marshall and L. M. Marx, *The Economics of Collusion: Cartels and Bidding Rings*, Cambridge, MA: MIT Press, 2012.

[3] M. C. Levenstein and V. Y. Suslow, "What Determines Cartel Success?" (2006) 44(1) *Journal of Economic Literature* 1.

[4] P. Cramton and J. Schwartz, "Collusive Bidding in the FCC Spectrum Auctions" (2002) 1(1) *Contributions to Economic Analysis & Policy* 1078, at www.cramton.umd.edu/papers2000-2004/cramton-schwartz-collusive-bidding.pdf.

[5] C. Hocepied and A. Held, "The Assignment of Spectrum and the EU State Aid Rules: The Case of the 4th 3G License Assignment in France," competition policy newsletter 2011-3, at http://ec.europa.eu/competition/publications/cpn/2011_3_6_en.pdf.

[6] *Ex parte* Submission of the United States Department of Justice: In the matter of Policies Regarding Mobile Spectrum Holdings. Before the Federal Communications Commission. WT Docket No. 12-269 (April 2013), 10–11.

[7] See Ofcom, "Assessment of Future Mobile Competition and Award of 800 MHz and 2.6 GHz," statement (July 24, 2012), at http://stakeholders.ofcom.org.uk/consultations/award-800mhz-2.6ghz/statement.

[8] M. Cave, "Anti-competitive Behaviour in Spectrum Markets: Analysis and Response" (2010) 34(5–6) *Telecommunications Policy* 251.

[9] P. Cramton, E. Kwerel, G. Rosston, and A. Skrzypacz, "Using Spectrum Auctions to Enhance Competition in Wireless Services" (2011) 54(4) *Journal of Law and Economics* S167.

[10] P. Cramton, "Why Spectrum Caps Matter." The Hill (February 18, 2014), at http://thehill.com/blogs/congress-blog/technology/198623-why-spectrum-caps-matter.

[11] M. Cave and W. Webb, "Spectrum Limits and Auction Revenue: The European Experience" (April 2013), at http://apps.fcc.gov/ecfs/document/view?id=7520934210.

[12] G. Myers, "Spectrum Floors in the UK 4G Auction: An Innovation in Regulatory Design," at www.lse.ac.uk/researchAndExpertise/units/CARR/pdf/DPs/DP74-Geoffrey-Myers.pdf.

[13] A. Bavasso and D. Long, "The Application of Competition Law in the Communications and Media Sectors: A Survey of Recent Cases" (2014) 5(4) *Journal of European Competition Law & Practice* 233.

[14] See http://europa.eu/rapid/press-release_MEMO-14–387_en.htm.
[15] BEREC/RSPG on infrastructure and spectrum sharing in mobile/wireless networks (June 16, 2011), available at http://rspg-spectrum.eu/wp-content/uploads/2013/05/rspg11-374_final_joint_rspg_berec_report.pdf.
[16] P. Milgrom, J. Levin, and A. Eilat, "The Case for Unlicensed Spectrum" (2011), available at http://web.stanford.edu/~jdlevin/Papers/UnlicensedSpectrum.pdf.
[17] Indepen, Aegis & Ovum, "The Economic Value of Licence Exempt Spectrum" (January 2008), at www.ofcom.org.uk/research/technology/overview/ese/econassess/value.pdf;
[18] R. Katz, "Assessment of the Economic Value of Unlicensed Spectrum in the United States" (2014), at www.wififorward.org/wp-content/uploads/2014/01/Value-of-Unlicensed-Spectrum-to-the-US-Economy-Full-Report.pdf.
[19] M. Bykowski, M. Olson, and W. W. Sharkey, "A Market-Based Approach to Establishing Licensing Rules: Licensed versus Unlicensed Use of Spectrum" (2008), OSP Working Paper Series, FCC.

6 Spectrum trading

6.1 Introduction

Auctions are episodic affairs. But with innovation being a distinctive feature of the wireless communication industry, it is desirable to have ready access to suitable frequency bands more regularly available. Spectrum trading is the process that allows the transfer of certain rights to use radio frequency spectrum from one undertaking to another.[1] Smooth trading mechanisms enable the flow of spectrum resources among users and can contribute to providing speedier access to spectrum, compared to traditional administrative methods.

Historically, spectrum trades have not been possible between users entitled to rights on spectrum. Hence, in order to assign frequency bands to a different user, spectrum had to be returned to the spectrum manager and then reassigned – a much more rigid mechanism than secondary trading, with high transaction costs and little potential to inform on spectrum values [1].

The introduction of spectrum trading has a number of advantages. Spectrum trading makes it easier for prospective new market entrants to acquire spectrum and develop their business, it enables fast-growing companies to expand more quickly than would otherwise be the case, and it can provide considerable incentives for incumbents to invest in new technology in order to ward off the threat of new entrants. Therefore, spectrum trading promotes dynamic efficiency, as it encourages innovation, risk taking, and efficient use of inputs – including spectrum. Indeed, spectrum users can minimize costs by dynamically choosing the most efficient combination of spectrum and other production factors to deliver their goods and services.

For instance, after a primary spectrum assignment, either by an administrative procedure (e.g. "first-come, first-served") or by a market mechanism (an auction), the holder of spectrum rights may realize that the value being attached to those rights is lower than another undertaking's valuation. In this case, it would not be efficient to leave spectrum in the hands of the current spectrum holder, as efficiency is usually achieved when the users of spectrum tend to be those with the highest valuations for the spectrum. A trade will only take place if the spectrum is worth more to the new user than it was to the old user, reflecting the greater economic benefit the new user expects to derive from the acquired spectrum [2].

---

[1] It can also refer to leasing agreements, but, as we will illustrate below, spectrum leasing is much less used than transfer.

There are, however, a few aspects of secondary spectrum trading that have to be considered in order to exploit its potential, or that suggest caution. Implementation of spectrum trading requires a coherent framework, where, for instance, tradable spectrum rights are clearly specified and trading markets are functioning well. To facilitate trades, it is crucial to establish a swift and inexpensive mechanism, with transaction costs as low as possible (otherwise, if transaction costs are too high compared to potential efficiency gains, these gains will not be realized). However, spectrum trades should not dampen competition.

In liberalizing countries, spectrum regulators have been introducing market-based mechanisms for spectrum management. Those mechanisms include change of assignment (by trading) of spectrum used for commercial services. But, compared to the introduction of other market mechanisms – notably auctions – spectrum trading has not been implemented or used as widely. Only some frequency bands have tradable spectrum rights attached to them, while frequency bands used by the public sector are substantially managed in the traditional way. Furthermore, regulators have considered possible anticompetitive behavior in trading and favored a cautious approach in newly created spectrum markets.

The aim of this chapter is to illustrate how spectrum trading can make spectrum management more efficient and how safeguards can be put in place to avoid trading abuses (i.e. those which might harm consumers). Experience in spectrum trading in a number of liberalizing countries is also briefly presented and discussed.

6.2 Spectrum secondary markets

Spectrum trading plays an important part in securing optimal use of spectrum by relying on markets, because this enables spectrum to migrate to users that will use it most efficiently. However, a number of conditions must be satisfied to achieve full efficiency of spectrum markets [3].

First, tradable spectrum rights need to be clearly defined; indeed, "the scope for successful secondary-market transactions is inextricably linked to the initially specified usage and property rights" [4, p. 62]. Second, information about those rights should be readily available. Market forces are more effective when traders are better informed about traded goods and valuations. Spectrum registries can lower transaction costs and facilitate trading. In fact, international organizations and individual countries, especially liberalizing ones, have been establishing and updating spectrum registries to make information available.[2]

In the case of spectrum, interference is a fundamental issue and readily available information on spectrum rights and transfers is crucial, especially to guarantee

---

[2] In Europe, Commission Decision 2007/344/EC, on harmonized availability of information regarding spectrum use within the Community, has promoted the availability of information on the use of radio spectrum through a common information point and by the harmonization of format and content of such information.

protection from harmful interference. Full information enables market forces to accurately take all costs into account – including interference costs – thus avoiding unforeseen externalities. Market forces can therefore steer prices more effectively towards efficient levels.

In fully efficient spectrum markets, none of the traders has market power – otherwise a trader would be able to affect unilaterally the operations of the market and, consequently, market prices. This can be more easily achieved if markets are thick, i.e. if there are many buyers and sellers, as market prices can more accurately reflect underlying spectrum value [5]. In thick markets, price signals, which typically embody a lower "spread" between the buy and the sell price, stimulate more efficient trading. Hence, spectrum can be more easily allocated to the most valuable uses. Furthermore, spectrum brokers or band managers can play an important role in matching spectrum demand with spectrum supply, thus facilitating trades.[3]

Finally, departure from command-and-control regulation towards a market-based approach to spectrum transfers has to be matched with the development of an effective dispute resolution process. Legislation cannot be far-reaching in the specification of actual arrangements, as the amount of (detailed) information, which has to be provided to achieve an agreement on a spectrum transfer, can be huge. Therefore, a resolution process would arbitrate on problems arising from transgressions of interference rights and responsibilities by one party or another [6].

6.3 Forms of spectrum trading

Spectrum trading includes various types of transaction. The most relevant type consists of spectrum transfers, which can be either total or partial spectrum transfers, as discussed below. Spectrum transfers differ from spectrum leasing because the latter type would be time-dependent, whereas in a spectrum transfer the rights (and associated conditions) are transferred for the remaining duration of a license. Some types of transfer do not represent true secondary-market transactions: relevant cases occur when licenses are transferred among affiliates of the same parent company, or when a license holder is acquired by, or merged with, another firm – possibly because of its spectrum license holdings [4].

Spectrum transfers involve negotiations between two (or more) undertakings. For total transfers, a license is transferred in full and all the rights and conditions attaching to a right of use of spectrum are transferred from one party to another. However, transfers can be partial. This may occur either because the transfer involves only a portion of licensed spectrum – although all the rights and obligations attaching to that portion of spectrum are transferred – or because some of the license conditions would not remain applicable to the spectrum transferred to another undertaking.

A split of rights or change of conditions governing previously licensed spectrum would normally trigger a regulator's requirement to analyse the proposed variations.

---

[3] For instance, in the US, this is one of the aims of Spectrum Bridge (see spectrumbridge.com).

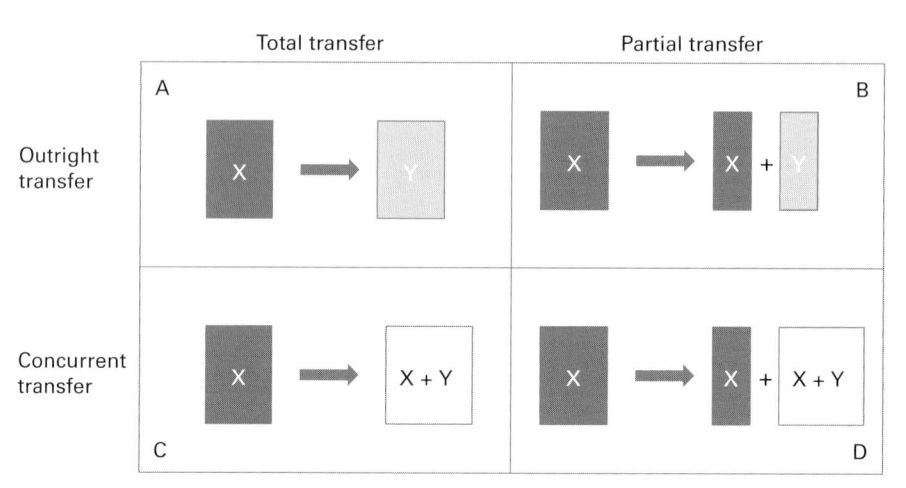

Figure 6-1. Types of transfer. Source: [7], p. 4.

This assessment is usually preliminary to issuing a new license to the transferee, while the original license holder would continue to be required to meet all the conditions of its license. Regulatory scrutiny aims at assessing whether the proposed variations are in line with spectrum regulations and would not effectively place other parties, who acquired similar rights of use over spectrum, at a disadvantage.

Transfers of rights can be either total or partial, and both total and partial transfers can each be either outright or concurrent transfers, depending on whether the transferor shares any rights and obligations with the transferee afterwards. In outright transfers, all the rights and obligations under the license transfer from one party to another; whereas in concurrent transfers, the transferred rights and obligations become rights and obligations of the transferee, while continuing to be rights and obligations of the transferor. To illustrate, Figure 6-1 considers two licensees (X and Y). In Box A, firm X transfers all his rights and obligations to firm Y and retains none of them. In Box B, X trades with Y and retains some of the original rights and obligations, but does not share any of them with the new licensee Y. In Box C, the new licensee receives X's rights and obligations, but these are shared with the transferor (X). Finally, in Box D, the original licensee, X, transfers only a few of his rights and obligations, and the new licensee Y acquires these rights and obligations on a shared basis with X.

Spectrum transfers are negotiated for the remaining duration of the license. However, parties may want to transfer spectrum for a shorter period of time. In those circumstances, they may prefer to use another type of transfer, i.e. spectrum leasing. In a lease situation, transfers of spectrum rights and obligations are time-limited. But only few jurisdictions have established frameworks for spectrum leasing (see, for instance, the US and European cases in section 6.5 below). By and large, spectrum leases, as well as transfers, must be notified to national regulators. Regulators are lifting administrative burdens and therefore smoothing spectrum trading. This may involve a different regulatory approach, especially with regard to assessments of possible competition distortions.

6.4 Competition concerns and other objections to spectrum trading

Maximizing the opportunities for spectrum-using industries requires that spectrum be fully used rather than hoarded, and that no firm is able to abuse dominance in the market for spectrum itself or to foreclose or limit competition in end-user markets. The development in recent years of the use of market methods (to allocate and assign spectrum in place of more traditional administrative methods) has focussed attention on the risks of anticompetitive conduct in the newly created spectrum markets. Indeed, any change to the spectrum holdings of operators in a market could potentially distort competition in that market. This would affect competing firms as well as consumers. This section considers the assessment of anticompetitive behavior and its possible remedies. It also considers whether trading can have a negative impact on harmonization and the issue of windfall gains.

6.4.1 *Ex ante* versus *ex post* assessment of anticompetitive trades

It is the regulator's duty to ensure that spectrum trades do not distort competition.[4] Regulators have a choice between two different approaches to dealing with competition concerns. On the one hand, they can choose an *ex post* spectrum trade assessment approach, i.e. spectrum trades do not require prior regulatory approval – nevertheless, implemented trades are subject to regulatory intervention and possibly to being reversed if competition is negatively affected. On the other hand, regulators can choose an *ex ante* assessment approach, i.e. spectrum trades must be preliminarily authorized by the regulator.

Both approaches have pros and cons, and very different implications. Moreover, according to circumstances, regulators might run into either so-called type 1 or type 2 errors. Type 1 errors occur when a proposed trade is subjected to an *ex ante* assessment of its likely effects upon competition, but the regulator's test is too restrictive and so the regulator decides to prevent a trade which would not have distorted competition. On the other hand, type 2 errors occur when a proposed trading was not prevented, even if it was likely to be anticompetitive, as the regulator's test was not sufficiently demanding.

Compared to an *ex ante* assessment of the likely competitive impact of a proposed trade, an *ex post* review of notified spectrum trades can lead to faster trade execution. Therefore, a higher level of trading may be encouraged. This may be favored by operators and potential new entrants, who can also benefit from lower administrative burdens and compliance costs for notifying parties, at least at the outset of any trade. However, a post-trade assessment review offers less protection to those not involved in the trade itself. In particular, consumers may be negatively affected (and for a considerable period of time). Furthermore, a post-trade assessment approach may be favored by an operator that was seeking to engage in a trade that could have anticompetitive effects, as it may be hard to isolate the effect of the trade from other market developments in

---

[4] Any trade has to be registered to be effected. In most jurisdictions spectrum trades are not subject to the normal mergers regime, unless they arise as a result of a merger of "enterprises."

> **Box 6-1.** *Ex ante* versus *ex post* spectrum trade assessment
>
> A. *Ex ante* assessment
>
> **Pros:**
> - spectrum trades must be preliminarily authorized by the regulator,
> - parties not involved in a trade might be protected better by a prior approval.
>
> **Cons:**
> - the regulatory burden is time-consuming and costly,
> - the regulator might prevent a trade which would not have distorted competition.
>
> B. *Ex post* assessment
>
> **Pros:**
> - trades can be implemented faster and with less administrative burden,
> - benefits from spectrum trading can be accrued earlier.
>
> **Cons:**
> - the regulator might not prevent an anticompetitive trade,
> - it may be difficult to isolate the anticompetitive effects of a trade from other market developments.

some *ex post* assessment investigations. One reason is that it may be difficult to establish that any perceived distortion in competition is attributable to a spectrum trade which has already occurred; another reason is that it may be especially difficult to convince a court that a spectrum trade should be dissolved [8, 9].

These drawbacks of an *ex post* review of spectrum trades can lead regulators to prefer an *ex ante* approach. There are, however, possible inconveniencies with prior assessment scrutiny of proposed trades. The major drawback is that regulatory actions of those kinds are more time-consuming and costly. Therefore, the benefits of "pro-competition" trades would be delayed for firms and consumers. Moreover, the regulator would determine whether a notified trade may or may not be put into effect, or may be put into effect subject to some specified conditions. This could discourage trading.

6.4.2 Possible remedies to competition concerns

Spectrum regulators can rely on various remedies to avoid anticompetitive behavior in spectrum markets. Some of these embrace the prior-assessment approach, e.g. spectrum caps and sector-specific interventions to control secondary trade. Remedies such as "use-it-or-lose-it" clauses and application of competition law are instances of post-trade assessment regulatory measures. We will now briefly review those remedies.

Section 5.5 discussed the applications of spectrum caps in auctions, where they form part of the auction rules. They can apply either to the amount of spectrum acquired in a particular award or to a firm's total holdings after the auction. But caps of the latter sort can also restrict a firm's (or a group's) trades. The best known spectrum cap was that imposed on the United States commercial mobile radio spectrum between 1994 and 2003. That cap placed a limit of 45 MHz (raised to 55 MHz in 2001) on the spectrum which a single entity could acquire in any geographical area. Spectrum caps have also been proposed or implemented in other American countries, such as Canada, Guatemala, and Mexico, but have not been used in Australia, where they were decisively rejected by the Productivity Commission in its 2002 review of spectrum policy [10].

The problem with spectrum caps is that they might distort competition by preventing a firm increasing its spectrum holdings even when it would be efficient to do so [11]. A firm providing its customers with innovative services of high quality at low prices can virtuously gain market shares. Moreover, if, as is likely, there are economies of scope in providing services, existing operators might be best placed to meet the new demand. In those circumstances, if ("hard") caps were strictly used by regulators, a firm might find itself either unable to meet demand, or only able to do so by a costly reuse of frequencies, which would be inefficient and would not be in end users' interests. To deal with drawbacks of this kind, caps could be expressed in terms of a percentage of spectrum in use to supply a particular group of services, or there could be fairly frequent reviews of the cap. In this way, "soft" caps could provide a better remedy. Under a "soft" cap, exceeding the spectrum quota simply triggers a license condition, which might, for example, entitle the regulator to undertake an investigation and require divestment of spectrum if it is not satisfied either that there were no competition problems or that they were being addressed.

Where caps can be waived or are "soft" caps, the boundary between vetting of trades and use of caps becomes imprecise. As noted above, a spectrum regulator can operate in a framework of sector-specific legislation which gives it the power to approve or disapprove any trade.[5] This would essentially involve establishing whether the trade led to a "significant lessening of competition" (SLC).[6]

For the introduction of "use-it-or-lose-it" license clauses, the license contains a condition according to which a penalty (e.g. a fine, or surrender of unused spectrum) is enforced if the spectrum is not used. There are a variety of problems with this approach. First, there may be good reasons – and not anticompetitive ones – to acquire spectrum ahead of use. Having guaranteed access to spectrum may be a precondition for making more substantial investments in technological development or equipment needed to provide the service. Moreover, newcomers to the industry might be discouraged from entering markets, especially because they would be unable seamlessly to switch spectrum from a previous to an innovative use. Second, where spectrum usage

---

[5] For instance, in Europe, the legal position varies from member state to member state. The general rule is that mergers take place between undertakings, which means that the acquisition of a spectrum license – which is not an undertaking – is not subject to European merger regulations.

[6] The SLC test is used in a number of countries, including the UK, the Netherlands, Poland, and Australia (see [9]).

data are monitored, they often show low levels of utilization. But in order to enforce a "use-it-or-lose-it" license condition, efficient episodic utilization would have to be distinguished from authentic anticompetitive behavior.

6.4.3 Other concerns

There are two other sources of concern that will be briefly considered here. The first issue is harmonization and the second one is windfall gains. In evaluating use of market methods, losing the benefits of harmonization (especially economies of scale, if any) is a factor to consider. There is no clear evidence suggesting that a market-based approach is preferable to the administrative alternative. In addition, not all countries are adopting *carte blanche* trading (e.g. some countries permit trading only for spectrum sold at auctions and at least pre-notification of trades is still commonly required).

The second concern is about windfall gains, which accrue to owners of specific property rights, without any effort or economic activity. There would only ever be cause for concern if excessive profits can be made without taking on correspondingly higher risks, simply by trading without engaging in any productive activity, or if spectrum had been obtained using non-market-based mechanisms. It is possible that trading, especially where change of spectrum use is allowed, may raise concerns of windfall gains. However, this should not be an impediment to secondary spectrum markets, and usual instruments (such as taxation) can be relied upon where appropriate. Moreover, empirical research suggests that, on the one hand, more liberal license rules, including possible change of spectrum use on transfers, tend to increase values of spectrum in growing demand; on the other hand, liberal licenses can ultimately reduce artificial spectrum scarcity, lower spectrum values, stimulate productivity and competition, and reduce excess profits arising from scarcity [12, 4].

6.5 Spectrum trading in practice

Several countries in the forefront of spectrum management reform have been introducing market mechanisms for spectrum assignment. With regard to secondary trading, this section provides a brief overview of spectrum policy developments in Europe (with more focus on spectrum trading in the United Kingdom), the United States, Central American countries, Australia, and New Zealand. Other liberalizing countries (e.g. India, Canada, Nigeria, and Japan) have taken action to promote market-based assignment of spectrum, but by and large those actions have been limited to the introduction of spectrum auctions. For instance, Canadian licenses allocated through auctions are tradable, but secondary trading is dormant. In India, the regulator has recently opened up the way to introducing trading of licenses assigned by auction, whereas a 2005 review had recommended not allowing trading at that time.

6.5.1 European countries

Most European countries have implemented spectrum trading in a number of bands used for commercial services; however, frequency bands open to spectrum trading differ among countries. In Denmark, trading of spectrum usage rights has been allowed since 1997; in Switzerland, trading was introduced in the following year. Article 9 of the Framework Directive 2002/21/EC permits EU member states to allow for the transfer or lease of rights to use radio frequencies between undertakings.[7] Therefore, spectrum trading and leasing is allowable, but only mandatory in certain major mobile bands. In 2004, the RSPG (Radio Spectrum Policy Group) adopted a cautious stance on trading and considered it to be beneficial in certain parts of the spectrum,[8] subject to the implementation of sufficient safeguards to ensure that potential benefits are not offset by adverse consequences. The RSPG favored a phased approach to secondary trading of rights of use to the spectrum, leaving to individual countries the decision whether to introduce secondary trading and the timing of it. This took into account that some EU countries were introducing secondary trading (e.g. Austria, Sweden, and the UK), while other countries were more hesitant [13].

The 2009 revised Framework does not impose pre-notification and consent conditions on any leasing schemes. This should promote more simple national trading regulations, encourage spectrum trading, and allow for the development of new trading processes.[9]

In 2011, the ECC reported that, in the CEPT geographical area, only four countries had declared that they do not allow trading of usage rights. Moreover, spectrum trading is permitted under circumstances which differ from country to country. For instance, in Austria all frequencies which are defined as scarce, or assigned by auction, are tradable; whereas in Lithuania scarce spectrum is not tradable.

European countries exhibit a variety of trading options and processes. As frequencies, geographical areas and the duration of a license can be subdivided, usage rights in frequency or geography can be traded in some countries (e.g. Czech Republic, Denmark, Norway, Spain, and the UK), whereas in other countries they can be traded neither in frequency nor in geography (e.g. Croatia and Romania). A few countries permit trading in time (e.g. Austria, France, and Spain). Moreover, Denmark and the UK have spectrum leasing by time. With regard to procedure, intent to trade must be notified. However, information required at notification, information to be published prior to the transaction, and approval procedures are rather heterogeneous [14].

In the UK, spectrum trading was introduced at the end of 2004 as a key element in Ofcom's program of market-based reform. Since then, trading has been progressively extended to a broad range of licenses. Under the Trading Regulations, which give effect to Article 9 of the Framework Directive 2002/21/EC, Ofcom has introduced trading

---

[7] The revised Framework Directive contains a number of decisions that facilitate EU member states to develop more flexible trading systems.
[8] The RSPG was sceptical about the application of trading in bands catering for government services (e.g. defense) and safety-of-life services (e.g. civil aviation), terrestrial broadcasting services, and broadcasting-satellite services, as well as scientific services (e.g. radio astronomy).
[9] See Directive 2009/140/EC of the European Parliament and of the Council of November 25, 2009, Article 9(b).

options which offer flexibility to parties interested in trading rights: in addition to an outright total transfer, the regulations permit concurrent or partial transfers.

Trading volumes in the UK have been low since trading was first permitted, particularly at the beginning. The Transfer Notification Register (TNR), which provides information on licenses traded or in the process of being traded, shows information on hundreds of trades.[10] However, only a few extra-group trades had been accomplished. This has been a source of concern to the regulator.[11] Furthermore, Ofcom has recently conducted work on simplifying spectrum trading in the UK, as features of its trading regime may be imposing unnecessary regulatory burdens [7].

Outside the UK, trading volumes appear even lower, especially where frequency bands other than private mobile radio (PMR) bands are taken into consideration.[12] In fact, the ECC report shows that, for PMR licenses, there were approximately 150 trades per year in Norway, 120 trades between 2007 and 2009 in Portugal, 69 trades in Lithuania, and 27 transactions in Romania. In fixed wireless access bands, there were 45 trades in France, and three trades in Denmark. There were three trades in mobile bands (two trades in Denmark, and one trade in Slovenia). The numbers reported are even lower in other bands (e.g. for PMSE and satellite licenses).

6.5.2 American countries

Moving away from Europe, in the US it has long been recognized that secondary markets can potentially serve as at least a partial correction to misallocation of spectrum, thus facilitating the flow of spectrum in the hands of those who value it the most [16]. The US secondary market has been the most active one across liberalizing countries, with thousands of trades made each year, facilitated through a trading marketplace and database with license ownership information – although the FCC does not record prices [17].

FCC procedures for spectrum leasing were substantially liberalized by the First Report and Order in October 2003, which provided two modes of liberalized arrangement. The first mode is spectrum manager licensing, where the licensee retains both de jure control and effective de facto control over the leased spectrum. The ruling enabled leases *without* prior FCC approval within the perimeter drawn by a license: in this mode, it is the licensee that is primarily accountable to the FCC for compliance with spectrum-relevant legal and regulatory obligations. The second mode is the de facto transfer mode, where the licensee retains de jure control. Although this is a fast-track approval process, prior FCC approval is still required. Spectrum leasing and subleasing are still stymied by the pre-notification requirement in the US. However, the Second Report and Order of 2004 further liberalized the process and made overnight processing of lease applications available to a wide variety of lease arrangements (the parties certify that the arrangement

---

[10] Access to the UK TNR is available at http://spectruminfo.ofcom.org.uk. Most trades involved business and radio licenses.
[11] "Even so, Britain appears to be light years ahead of France and Germany, which introduced the system around the same time" [15].
[12] PMR bands underpin such services as those used by local taxi firms.

does not raise any of a specified list of potential concerns, such as foreign ownership, license eligibility, or competition issues).

In Central America, Guatemala and El Salvador provide a somewhat different case of spectrum trading. The Salvadoran regime, while technically different from Guatemala's, yields a similar set of spectrum rights [18]. After 1996, usufructs over spectrum (*titulos de usofructos de frecuencias*, or TUFs) were introduced in Guatemala, where the law permits change of use on transfer. The sale of a TUF is accomplished by its endorsement by the seller, the buyer registering its new rights with the independent spectrum body. El Salvador's law does not require a spectrum registry.

6.5.3 Australia and New Zealand

Australia was one of the very earliest countries to allow spectrum trading. First, apparatus licenses became tradable in 1995.[13] Second, spectrum licenses, which were introduced in 1992 and then issued in 1997, are Australia's most market-oriented licenses, and they are tradable.

In Australia, spectrum blocks owned by licensees are represented in standard trading units (STUs), which are the smallest spectrum units recognized by the regulator. STUs can be combined vertically, to provide increased bandwidth, or horizontally, to cover a larger area. Notwithstanding the introduction of STUs, the rate of trading has been quite slow [19]. Moreover, the Productivity Commission (PC) itself noted that many transfers were among related parties.[14]

Recently, the Australian Communications and Media Authority (ACMA) has been implementing a number of measures to remove regulatory barriers to spectrum trading and leasing, as well as to provide better information to users, especially following a review of spectrum trading in late 2008. As a result of this review, the ACMA (i) has reduced regulatory burdens and encouraged spectrum trading in various ways, e.g. by introducing an online system (to replace the paper-based system for registering license trades); (ii) has commenced a voluntary registration of third-party authorizations, with a view to providing the market with more information; and (iii) has developed a trading page on the ACMA website to provide the market with better information on license trades and transfers [19].

In New Zealand, the Ministry of Economic Development (MED) published a review of radio spectrum policy in 2005, which noted that the level of trading had been low and mainly confined to FM and AM radio broadcasting licenses (where a great deal of consolidation happened through takeover). In addition, trades had not involved a change in use. The small size of the market in New Zealand, entry barriers to sectors using radio spectrum and availability of alternative spectrum were identified as factors limiting

---

[13] The Australian Radiocommunications Act 1992 provided for a new, comprehensive system of licensing. A *spectrum license* represents the more market-oriented form of licensing; in particular, it is fully tradable and is issued for periods of up to 15 years. An *apparatus license* (the traditional command-and-control-type license) generally authorizes the operation of a transmitter or receiver at a particular location.

[14] The PC also identified a number of possible reasons for the slow supply of spectrum traded in secondary markets [20].

secondary spectrum trading. To facilitate trades, in 2009 the MED completed a major technology platform upgrade to the online public register of radio frequencies.

6.5.4 Recent policy developments

Before moving to a few concluding remarks, this section briefly discusses recent developments in spectrum management methods which may have an impact on spectrum markets. One such development is access to TV white space spectrum (TVWS), or, as Ofcom calls it, interleaved spectrum. See Chapter 9 below.

Even though the trend has been towards unlicensed exploitation of TVWS, some research has suggested that there are opportunities to exploit licensed TVWS, and to introduce trading there. Indeed, the FCC has approved the licensed white spaces. Furthermore, recent research suggests that the potential of cognitive radio may be more fully realized in licensed white space, and that this would be encouraged by the emergence and maturation of rules for spectrum transfer – perhaps arranged by a band manager. Along these lines of research, Bogucka et al. [21] studied the case of a spectrum broker that manages a TVWS secondary spectrum market.[15]

Taking advantage of TVWS and cognitive radio developments, the European COGEU project also promotes the introduction of real-time secondary spectrum trading. The innovation brought by COGEU is the combination of cognitive access to TVWS with secondary spectrum trading mechanisms [22].

In the US, the National Science Foundation is running a project to develop S-Trade, an auction-driven spectrum trading platform to implement the spectrum marketplace. S-Trade would serve many small players and enable on-the-fly spectrum transactions, by selectively buying idle spectrum pieces from providers and selling them to a large number of spectrum users [23].

Although much European spectrum management is still rooted in the command-and-control tradition, recent spectrum-sharing discussions in Europe envisage the introduction of additional licensed spectrum users into already licensed bands, under the new Licensed Shared Access (LSA) concept [24]. LSA is a complementary spectrum management tool aimed at facilitating the introduction of new users in a frequency band, while maintaining private and public incumbent spectrum users in that band.[16] These new users, e.g. mobile network operator (MNO) LSA licensees, would share the spectrum with the incumbents under the supervision of the regulator. A key aspect in the implementation of the LSA concept is the development of a set of sharing rules and conditions, which may lay the basis for developing a more market-based trading approach to LSA. There has been some debate about whether such shared rights should be traded or whether they should have attached to them administrative prices of the kind considered in Chapter 7 below [25].

---

[15] A real-world test scenario has been successfully applied in Germany by the research team. The spectrum broker may operate in a merchant mode (i.e. spectrum base price is decided by the allocation procedure) or an auction mode (i.e. spectrum final price is decided by bids).

[16] The current studies are looking at the 2.3–2.4 GHz band, where there are different incumbents in different countries, such as program making and special equipment (PMSE) in Finland and the military in France.

6.6 Concluding remarks

Secondary spectrum trading is a means to achieve more efficient use of radio frequencies. It complements the introduction of market-based mechanisms for primary assignment of spectrum, and it is arguably a more potent mechanism than auctions, for the reason that the opportunities for efficiency-enhancing exchanges are always in place.

Yet in practice its impact has been modest so far. In liberalizing countries, the level of trading has been generally lower than expected, at least outside the US. Several possible reasons have been suggested, with each likely to have had some influence [3, 26, 27]. Some of those reasons are closely related to features of spectrum markets: for instance, commentators have considered insufficient information, inadequate development of private band managers, insufficient market liquidity, impediments from high transaction costs, and the difficulty of synchronizing the purchase and sale of spectrum for a firm which wished to switch bands.

A second set of reasons is more closely related to the regulatory framework, such as uncertainties due to phased liberalization of spectrum use, lack of alignment of (tradable) license terms and conditions with firms' investment programs, strictness of regulatory conditions, and the extent of programs of primary awards. Other possible reasons lie between – for instance, the non-tradability of public spectrum holdings.

It might even be considered that, in a few countries, the initial assignment of frequencies has been efficient, and that no need to transfer spectrum has arisen afterwards. Nevertheless, spectrum trading can have only a small impact on spectrum efficiency, if unaccompanied by flexible spectrum rights enabling change of use. A few studies show that the benefits from liberalizing spectrum license conditions are much greater than the benefits from the trading of spectrum licenses on its own [1, 13]. But liberalization of license rules for changing spectrum use has been more complex, and its progress has been slower. Hence, national strategies of spectrum management reform may merit special consideration. Recent work analyses the complementarity between liberalizing spectrum use and permitting spectrum trading [28]. The analysis of spectrum policies further suggests that spectrum trading might be encouraged by increasing the flexibility of license use conditions. Therefore, spectrum regulators are, of course, in a position to keep on contributing to more efficient spectrum management, and, particularly, to the emerging of spectrum markets as a key component of spectrum assignment.

References

[1] Analysys, DotEcon, and Hogan & Hartson, "Study on Conditions and Options in Introducing Secondary Trading of Radio Spectrum in the European Community: Final Report for the European Commission" (2004), Cambridge.
[2] T. M. Valletti, "Spectrum Trading" (2001) 25(10–11) *Telecommunications Policy* 655.

[3] P. Crocioni, "Is Allowing Trading Enough? Making Secondary Markets in Spectrum Work" (2009) 33(8) *Telecommunications Policy* 451.
[4] J. W. Mayo and S. Wallsten, "Enabling Efficient Wireless Communications: The Role of Secondary Spectrum Markets" (2010) 22(1) *Information Economics and Policy* 61.
[5] C. E. Caicedo and M. B. H. Weiss, "The Viability of Spectrum Trading Markets" (2011) 49(3) *IEEE Communications Magazine* 46.
[6] G. R. Faulhaber, "The Question of Spectrum: Technology, Management, and Regime Change" (2005) 5(1) *Journal of Telecommunications and High Technology Law* 111.
[7] Ofcom, "Trading Guidance Notes" (2011), Doc. OfW513, London.
[8] ComReg (Commission for Communications Regulation), "Spectrum Trading in the Radio Spectrum Policy Programme (RSPP) Bands: A Framework and Guidelines for Spectrum Transfers in the RSPP Bands" (2012), Doc. 12/76, Dublin.
[9] Ofcom, "Ensuring Effective Competition Following the Introduction of Spectrum Trading: Statement" (2004), London.
[10] Productivity Commission, "Radiocommunications" (2002), Report no. 22, AusInfo, Canberra, Chapter 6.
[11] M. Cave, "Anti-competitive Behaviour in Spectrum Markets: Analysis and Response" (2010) 34(5) *Telecommunications Policy* 251.
[12] T. W. Hazlett, "Property Rights and Wireless License Values" (2008) 51(3) *Journal of Law and Economics* 563.
[13] RSPG (Radio Spectrum Policy Group), "Opinion on Secondary Trading of Rights to Use Radio Spectrum" (2004), RSPG04-54, Brussels.
[14] ECC (Electronic Communications Committee), "Description of Practices Relative to Trading of Spectrum Rights of Use" (2011), Report 169.
[15] D. Standeford (December 11, 2014), at www.policytracker.com.
[16] FCC (Federal Communications Commission), "Spectrum Policy Task Force Report" (2002), ET Docket 02-135, Washington, DC.
[17] S. Wallsten, *Is There Really a Spectrum Crisis? Quantifying the Factors Affecting Spectrum Licence Value*, Washington, DC: Technology Policy Institute, 2013.
[18] T. W. Hazlett, G. Ibarguen, and W. Leighton, "Property Rights to Radio Spectrum in Guatemala and El Salvador: An Experiment in Liberalisation" (2007) 3(2) *Review of Law and Economics* 437.
[19] ACMA (Australian Communications and Media Authority), *Spectrum Trading: Consultation on Trading and Third Party Authorisations of Spectrum and Apparatus Licences* (2008), Canberra.
[20] See www.acma.gov.au/theACMA/acma-media-release-1202010–8-october-2010-acma-takes-steps-to-improve-spectrum-trading-arrangements.
[21] H. Bogucka, M. Parzy, P. Marques, J. Mwangoka, and T. Forde, "Secondary Spectrum Trading in TV White Spaces" (2012) 50(11) *IEEE Communications Magazine* 121.
[22] C. Dosch, J. Kubasik, and C. F. M. Silva, "TVWS Policies to Enable Efficient Spectrum Sharing" (2011), 22nd European Regional ITS Conference, Sept. 18–21, Budapest.
[23] See www.nsf.gov/awardsearch/simpleSearchResult?queryText=0915699.

[24] ECC (Electronic Communications Committee), "Licensed Shared Access (LSA)" (2014), Report 205.
[25] E. Bohlin and G. Pogorel, "Valuation and Pricing of Licensed Shared Access: Next Generation Pricing for Next Generation Spectrum Access" (2014), working paper, Telecom ParisTech.
[26] OECD (Organisation for Economic Co-operation and Development), "Secondary Markets for Spectrum: Policy Issues" (2005), DSTI/ICCP/TISP(2004)11/FINAL, Paris.
[27] M. Weiss, "Secondary Use of Spectrum: A Survey of the Issues" (2006) 8(2) *Info* 74.
[28] L. F. Minervini, "Spectrum Management Reform: Rethinking Practices" (2013) 38(2) *Telecommunications Policy* 136.

7 Spectrum pricing and valuation

7.1 Introduction

This chapter is devoted to discussing how it is possible to derive a price or valuation of spectrum by means of a calculation rather than by implementing a market process. There are a number of circumstances in which this might be required:

- Suppose a mobile operator has been awarded spectrum by an administrative process, such as a beauty contest, in circumstances where access to the spectrum places the operator in a position where it can earn an excess return. The government may want to appropriate some of that excess return as revenue for the state, to finance necessary public expenditure. Charging the operator a fee for access to the spectrum may achieve that end.
- Suppose that public bodies have been assigned spectrum in the past, at no cost. As spectrum scarcity has grown, they will be sitting on an asset with valuable alternative uses, yet have little incentive to economize on it and return unneeded amounts for refarming. Requiring them to pay an annual fee may create an incentive to give unused spectrum back.
- Suppose the government or regulator intends to conduct an auction for a mobile band, and is concerned that operators may collude in the bidding process to keep prices down. One way of preventing such behavior from reducing revenues too much is to set a reserve price – an amount which the auction has to realize before the spectrum is licensed to the highest bidders. In this case, the government will want to set a price which prevents bidders from colluding but is not so high as to stop the sale.

In all these cases, a price is needed which does not come from implementing a sale process in a marketplace, but is calculated according to a different formula or procedure. It will transpire, from the examples above, there are several ways in which so-called "administrative" (as opposed to "market") prices can be set, differing in how the components in the price formula are established.[1]

---

[1] As well as market and administrative prices for spectrum, there is a third option – shadow prices. These play some role in determining the allocation of spectrum, even if they are not actually paid. For example, suppose a defense department is choosing between two weapons systems with the same capability, so that the choice depends on cost. In comparing the costs of the two systems, it is quite possible to include the notional or "shadow" price of the spectrum the two systems use, even if the department does not have to pay the prices.

Throughout the chapter, sometimes the discussion will focus on the annual price which, for example, a mobile operator may have to pay every year for a license granting access to a band. Sometimes it will be on a sum covering many years' access, for example a one-off up-front fee (such as is paid by someone buying a 50-year lease for a house) which provides access to spectrum for many years, or even in perpetuity. This sum is sometimes called a "valuation" of the asset.

The annual price and the multi-annual valuation are linked by a simple mathematical formula. Thus the obligation to pay an annual amount for 20 years can be re-expressed in terms of a single and immediate payment by the process of establishing the net present value of the sequence of annual payments. Equally, a single 20-year value can be converted into an equivalent stream of annual payments which can either be equal or can be tilted in some way to generate the same net present value.

These approaches to pricing can be distinguished from a system for making all spectrum users pay charges which are designed *not* to reflect the scarcity value of the relevant bands, but at least to cover in aggregate the costs of the spectrum management process. Historically, where spectrum is allocated administratively by command and control (that is, not via a price or auction method), such charges have been set, but typically at a level which is far too low to reflect the economic value of the more significant bands. Box 7-1 shows two fairly representative examples of how cost-recovery spectrum charges of this kind have been set in the recent past. They typically reflect government priorities or seek to avoid too acute a sense of unfairness among different groups of spectrum users, rather than to allocate spectrum efficiently.[2]

7.2   The separate components of spectrum prices

The standard "bottom-up" way to construct an administrative price or valuation for an input, product, or service is by summing the relevant components which are to be included. If the administrative price is to play the same role in spectrum allocation as a market price does in a market framework then it is instructive to examine the components which might go into an administrative price of this kind.

Starting from this angle, we first note that spectrum is a natural resource requiring no expenditure on extraction or transportation. Accordingly, its "cost" in use is not the cost of manufacture, but the opportunity cost of its use in its current application.[3] We take this to be the extra costs which would be incurred to produce the same output for which it is an input, if the quantity of spectrum in question were unavailable.[4]

A very important aspect of opportunity-cost prices follows from this: if the spectrum in question is in excess supply, its opportunity cost is zero. If a particular quantity of

---

[2] For further examples see [1].
[3] The notion of valuing an asset in accordance with its opportunity cost has been a feature of commercial and regulatory life for centuries. The concept was used by Benjamin Franklin in 1748 and the term was introduced in 1914 by Friedrich von Wieser in his book *The Theory of Social Economics*.
[4] We discuss in section 7.4 below another concept of opportunity cost, which is the value of the same spectrum in the production of an alternative output.

Box 7-1. Charges set to recover spectrum management costs: two examples

A. Thailand

Government departments do not pay charges and state enterprises pay lower rates than non-state enterprises. The last pay according to a formula:

$$FF = (BW \times AC \times FC) \times (N1 + 2N2 + 4N3)$$

where

FF = the frequency fee
BW = the bandwidth occupied
AC = the application constant
FC = the frequency constant
$N1$, $N2$ and $N3$ are numbers of terminals with different output power.

The frequency constant, FC, is much larger for lower than for higher frequencies:

Frequency	Frequency constant
0.01–1,000 MHz	10
1–3 GHz	5
3–10 GHz	0.5
10–20 GHz	0.05
> 20 GHz	0.001

The charging regime thus creates a differential of 10,000 to 1 between the price of spectrum below 1 GHz and above 20 GHz, and includes a use-specific application constant. (The former reflects relative scarcity, but not the latter.) A charge is also made for receivers.

B. Canada

Canada has developed a model in which spectrum consumption is recorded in three dimensions: bandwidth, geographic coverage, and exclusivity of use. Larger bandwidths, greater geographic coverage, and exclusive use of a spectrum assignment result in higher fees.

However, two licenses identical in these three dimensions may have widely divergent real values because of geographic location, spectrum in a major city presumably being more valuable than spectrum in the high Arctic, for example.

To account for these differences, and given the difficulties inherent in trying to determine true market values in the absence of a functioning market, the concept of spectrum scarcity has been applied as a sort of proxy variable. A grid/cell pattern has been overlaid on the geography of Canada, and in each cell, the volume of spectrum consumed by all users in a given band is divided by the total volume of spectrum existing in that band. It is this ratio that will determine the relative levels of fees across the country. In areas where spectrum use is high, such as major cities, the spectrum scarcity measure, and as a result the license fee, will also be high.

However, because the total of fees collected is capped to recover spectrum management costs, even though relative charges for different areas and bands may reflect relative scarcities, the overall level of charges fails to reflect the spectrum's true scarcity value. This means that bands are at serious risk of remaining in low-value uses.

spectrum were withdrawn, it could be replaced by another equivalent amount. Equally, any other user wanting access to that band could get its supply from the available excess. Setting a positive price would be detrimental: if spectrum in a band were a free good, attaching a price to it might take it out of use when it has no opportunity cost.

As we discuss below, one way of setting an administrative price for spectrum is simply to set it equal to its opportunity cost, as calculated in the following way: if the firm in question did not have access to that particular band, and had to replace it with another band, or with a non-spectrum input, what additional costs would it incur? Those extra costs measure the loss of the opportunity of using the spectrum in question. However, access to spectrum may confer other benefits as well. In a market context this can be appreciated by considering the thought processes which a firm is likely to go through in deciding how much to bid for a spectrum license lasting, say, twenty years.

The first factor, which we have already identified, is the extent to which holding the license will reduce the cost to the owner of producing the services the owner wants to produce, as against the next-best alternative. But there are two other factors:

- the extra profits the owner will make as a result of the market power that ownership of the spectrum license may confer on it, and,
- where a spectrum license allows change in use, or where it can be traded, or where it is subject to renewal, the "option value" available to the operator to convert the spectrum to a more advantageous use, to sell it at a profit, or to have an advantage in a renewal competition.

A firm's willingness to pay in the marketplace for a spectrum license will take account of all three of these components. It is likely to make the computation by projecting the revenues it will be able to generate by selling the spectrum-using service over the duration of the license (and any carryover there might be at renewal). From this it will subtract the other non-spectrum costs it must incur; including the cost of capital it has to pay to attract investible funds into its business. This procedure is sometimes called the "business plan" approach to spectrum valuation or "project pricing." The result is an estimate of the firm's maximum willingness to pay. Of course, the firm will try to acquire the license at a market price less than its maximum willingness to pay; the success it will have in doing this will depend in part upon the design of the auction.

Essentially an agency setting an administrative price can itself replicate this process, inserting its own estimated numbers into the business plan spreadsheet. The result of this calculation is the price setter's estimate of a firm's maximum willingness to pay. This obviously requires the price setter to have detailed knowledge of the costs, revenues, and competitive conditions of the sector.

An alternative way of setting an administrative price based on market principles for spectrum is to draw direct inferences from actual market prices paid in other bands or in other countries. This only works if the comparator bands or countries are sufficiently similar to allow the read across to the band and country for which an administrative price is being sought.

Thus there are broadly two bases on which administrative prices for spectrum are set. One only seeks to capture opportunity cost; the second captures other components of market price as well. Spectrum users on whom administrative prices are being imposed naturally have strong views about the level of prices they face. This was vividly illustrated when the United Kingdom government required the spectrum regulator Ofcom to change the basis on which administrative prices were set for certain mobile spectrum bands at 900 and 1,800 MHz, licenses for which were being renewed – from a relatively low opportunity-cost basis to the replication of full market prices observable in auctions. The initial increase was calculated at 475%. The UK mobile operators expressed outrage (the final increase is likely to be smaller) [2].

The next sections set out applications of the two pricing principles in more detail.

7.3 Finding opportunity-cost prices: an initial approach

One starting point for finding the opportunity cost of spectrum in a band is to set it equal to the additional costs which a firm would incur if it chose to produce the same outputs, replacing the original spectrum with the next cheapest set of inputs (which might be another band, or a non-spectrum input).

This corresponds roughly to what accountants call "deprival value." The term is apt because it expresses the loss the firm would suffer (expressed as an annual flow or as a capital sum over a run of years) if it no longer had access to the spectrum, but otherwise carried on.

The obvious way to estimate an opportunity cost of this kind is to compare the cost of producing a particular service using the spectrum band currently assigned to it with the cost of producing the same service using another technology. This will produce a number of estimates of cost savings, which can be expressed per megahertz of spectrum saved. In the absence of the band currently used, the firm would use the cheapest replacement, with the lowest additional cost per megahertz.

Take as an example the value of spectrum used for ubiquitous digital terrestrial broadcasting (DTT). Ways of delivering equivalent broadcast services include use of satellite broadcasting, the construction of a ubiquitous fixed network capable of providing the same service (for example, a cable or a fibre network), use of a transmission technology which economizes on spectrum but requires more sites (a so-called single-

Table 7-1. The opportunity cost of DTT spectrum in the UK

Alternative	Spectrum saving	Cost impact	Additional cost per MHz
1) Satellite transmission	48 MHz per multiplex	More dishes and set-top boxes	£60–80 million per MHz
2) SFN with more sites	< 40 MHz per multiplex	12 × as many sites	£7.5 million per MHz
3) SFN using DVB-T2	< 40 MHz per multiplex	New set-top boxes[7]	£0–25 million per MHz

frequency network or SFN[5]), and switching to a different transmission technology standard such as DVB-T2.[6]

By way of illustration, a report prepared for Ofcom by Plum Consulting has made illustrative estimates of some of these options, as shown in Table 7-1 [3]. In this instance, the lowest replacement cost is one of the latter two options, the lower value of which is somewhere in the £0–7.5 million range.

Before elucidating the concept of opportunity-cost pricing in more detail, it is worth making a number of points which flow from the discussion so far:

- The value per megahertz of a band is likely to vary with the increment of spectrum which is notionally being withdrawn. This can vary from a marginal amount (1 MHz) to a band as a whole (e.g. the 700 MHz band). If a marginal amount is withdrawn, it may be possible to work cheaply around its absence. In the example in Table 7-1, the notional withdrawal is of a major chunk of spectrum. The latter is the normal context in which spectrum pricing is done.
- As noted above, spectrum in excess supply has a zero opportunity cost; this flows from the fact that if a component of it were withdrawn, there would be zero cost of replacing it.
- Each firm with a different efficiency level is likely to face a different deprival value or opportunity cost. But since we are looking for a price which can be generally applied, it is better to do the calculation for a "representative firm."
- The "constant-output" approach is restrictive, since in practice different technologies will produce different quantities or qualities of output. For example, satellite distribution of broadcast programs may have more capacity than DTT. Or a mobile signal delivered on the 2.3 GHz band may penetrate buildings less successfully than the same

---

[5] An SFN uses the same frequency in adjacent cells to deliver the same broadcast content. If the radio interface is designed appropriately, the content in each cell is identical, and transmissions are synchronized, then this can result in significant spectrum savings.

[6] DVB-T2 (Digital Video Broadcasting – Terrestrial, version 2) is part of the DVB set of broadcast standards. It is an evolution of DVB-T which provides for greater spectrum efficiency through higher modulation modes and more efficient compression technology.

[7] Number of set-top boxes required depends on how many are already acquired.

Box 7-2. Taking account of uncertainty when setting administrative prices

Suppose a band which can be deployed to produce output 1 or output 2. As Table 7-2 shows, the central estimate of its value to produce output 1 is 100, but the value can lie with equal probability anywhere between 75 and 125. The central estimate of the value to produce output 2 is 75, falling with equal probability in the interval between 50 and 100. We assume risk-neutrality.

Table 7-2. Opportunity cost

		Estimates of opportunity cost of a band to produce two outputs					
		25	50	75	100	125	150
Output	Output 1				•———•		
	Output 2		•———•				

If we set a price of 100, there is no chance of the spectrum being used to produce output 2, and only a 50% chance of its being deployed in output 1, when the benefit there lies between 100 and 125. The expected benefit is accordingly low (in fact, 56.75). If we set a lower price, say 75, then it would certainly be of use to produce output 1, so the minimum value it would yield is 75. There is, however, a risk that the spectrum would be used to produce output 2, yielding a value between 75 and 100, when it could have a higher value, say in the range 100 to 125, producing output 1. But in this second case, the spectrum always yields some value, in no case less than 75, even if not necessarily the maximum. Thus in setting an administrative price in conditions of uncertainty, it makes sense to shade down the "best estimate" to avoid taking the spectrum wholly out of use by imposing too high a price.

signal transmitted on 800 MHz. In practice it may be inappropriate to focus exclusively on cost differences, fully excluding consumer benefits.
- There is no guarantee that a firm would be able to generate enough income to continue to provide the service if it were deprived of its spectrum. For example, digital radio (DAB or digital audio broadcasting) is normally rather fragile commercially. If it faced an opportunity-cost price for its spectrum, it might not be commercially viable. In the absence of a rival for its current allocation, charging a full opportunity price would be fruitless.[8]
- The ratio of the highest value to the lowest shown in Table 7-1 is fairly typical of opportunity-cost estimates. This makes it desirable to manage carefully the risk of setting prices too high. This is discussed in Box 7-2.

---

[8] The accountancy literature on deprival value notes this explicitly. Technically, deprival value equals the lower of replacement cost and recoverable amount, where the recoverable amount is the higher of net selling price and value in use. If the spectrum user cannot afford to replace the lost band, its value can simply be what it is worth to the current user (value in use).

- A problem can arise that the "next-best" option to a firm's current access to band A involves access to band B. To know the opportunity cost of band A we also have to know the opportunity cost of band B, which may itself require knowledge of the opportunity cost of band C, and so on. This introduces complexity, and even circularity, into the calculations, which may have to be tackled by a more comprehensive approach to spectrum pricing.

7.4 Interrelations among opportunity-cost estimates

The method of calculating opportunity cost described above assumes that the band being "priced" is efficiently assigned. It asks: how much extra would it cost to produce the output in question if the currently used, optimal band were not available? But it is equally pertinent to ask whether the current use is optimal. Table 7-3 illustrates this aspect of the problem.

The first row of the table shows the increase in cost (from the baseline) for the production of output X with band A, when the three inputs alternative to band A are band B, or band C, or the non-spectrum input D. The lowest such increase shown in the table is 10, which is taken as the opportunity cost of band A in production of output X.

But suppose that similar calculations show that the equivalent opportunity cost of band A in the production of output Y is higher. This is the case in Table 7-3, where the lowest increase in cost on the output Y baseline (the opportunity cost of band A in producing output Y) is 13. In other words, it looks as if band A is more usefully employed in producing output Y than in producing output X.

At first sight, this might be taken to mean that the "mistake" in assigning band A to output X should be corrected by switching it to produce output Y. But output X still has to be produced somehow. One possibility is to deploy the now-released band C, formerly used to make output Y, for this purpose. But it might turn out that use of this band is a very inefficient way of producing output X. In that case it might be necessary to make a whole chain of substitutions across the whole frequency table, each of which would require a new series of interdependent opportunity-cost calculations.

Thus with substitute bands, the opportunity-cost pricing problem is, in principle, one of "general equilibrium," in which efficient use of one band can depend on the pattern of

Table 7-3. Pricing example

Output produced	Band A	Band B	Band C	Non-spectrum input D
Output X	baseline for output 1	+10	+14	+15
Output Y	baseline for output 2	+16	+13	+20

Spectrum pricing and valuation

```
Incremental
unit cost of
mobile
service
provided on
various
frequencies

       900   1800  2300  3500
         Output of mobile services


Opportunity
cost of band
per MHz

       900   1800  2300  3500
              MHz in use
```

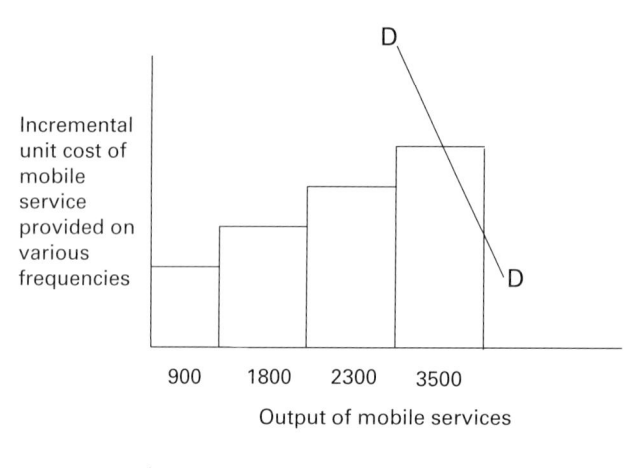

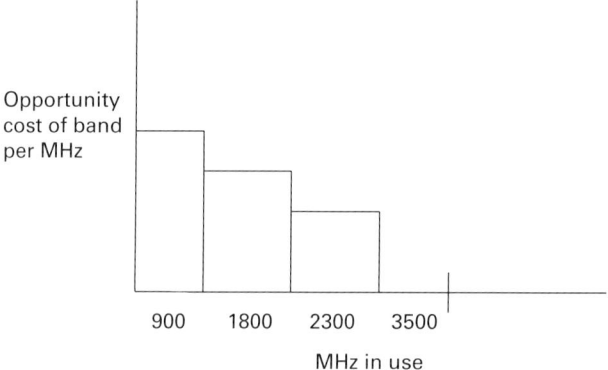

Figure 7-1. Relationship between incremental cost of radio access network and the opportunity cost of the spectrum at various frequencies.

use of any or all other bands. We are searching for a set of prices which together will support an efficient assignment of spectrum to produce the required set of outputs as a whole. With so many different bands available, and so many possible uses, this is computationally problematic, but probably not impossible. The harder task is to populate the very extensive set of cost data needed to implement the calculations.

From a practical point of view, it may be acceptable to adopt a simplified approach, if we are reasonably confident that current uses are optimal. One area where we may have such confidence is the pricing of mobile spectrum – simply because demand for such spectrum is growing so fast, with the rapid growth of mobile data services. Simplifying considerably, we can say that base stations working on lower frequencies provide mobile services at lower cost and possibly higher quality than base stations operating on higher frequencies. One reason for this is that signals at lower frequencies have a larger range; hence fewer base stations are required to deliver the same area of coverage. As a result, the graph of incremental costs (and, we assume, of prices), drawn against data traffic takes the form of the step function shown in the top part of Figure 7-1.

When demand is shown by *DD*, the marginal frequency, of which there is excess supply and hence an opportunity cost of zero, is the 3.5 GHz band. On that basis, the opportunity costs of the other frequencies are as shown in the lower part of Figure 7-1.[9] If higher demand had brought the next frequency into play, the 3.5 GHz band would come into the money, and the opportunity cost of all bands would rise.

To summarize, there are a number of pitfalls in pricing spectrum based on opportunity cost. The most obvious one is the difficulty of identifying and costing alternatives to the band in current use. Then there are complexities, which may not always arise, associated with the multiplicity of ways of using certain bands. It is sensible to manage the possibility of asymmetric errors in setting administrative prices by undertaking an appropriate risk analysis of the kind suggested in Box 7-2 above.

7.5 Opportunity-cost spectrum pricing in practice

Administered prices have been used most widely in the United Kingdom, where they are known as administered incentive prices or AIPs. Beginning in 1998, the then spectrum regulator began a staged introduction of such prices [4]. Initially, broadly two price bands for spectrum were set, a high price for spectrum deemed suitable for mobile use, and a lower price for "fixed" spectrum. This led to a rather sudden and arbitrary jump in prices – a "cliff face" – at a particular frequency [5].

The UK spectrum (and communications) regulator, Ofcom, adopted administrative pricing as a means of encouraging efficient use of spectrum for bands which were not auctioned, and began a program of extending administrative prices to additional uses, including public-sector uses.[10]

Ofcom's 2010 review of administrative pricing noted that administrative prices were by then widely applied, covering over 50% of spectrum [6]. The review reached a broadly favorable conclusion on its efficacy, made on the basis of general principles, supported by a rather limited and not entirely convincing evidence base [7], and identified general principles for future application shown in Box 7-3.

Considerable opposition can be expected from the firms and sectors asked to pay for spectrum charges. This applies generally, and does not exclude the noncommercial sector. Ofcom discovered this when it sought to impose charges on spectrum used for aeronautical and maritime purposes. Airlines asserted that it was a "tax" [8], which would undermine UK airports as hubs for travelers. A campaign against maritime spectrum charges was spearheaded by the charitable organizer of lifeboats. In the end, Ofcom was forced substantially to modify its original plans, eliminating charges for maritime search and rescue [9]. For radar and aeronautical navigation aids, Ofcom passed responsibility for strategic management of the spectrum to the government [10].

---

[9] Auction price data also support the view that spectrum in lower frequencies is worth more than spectrum in higher bands.

[10] See Chapter 12 for a more detailed discussion of public-sector spectrum management in the UK.

Box 7-3. Ofcom's 2010 administrative incentive pricing (AIP) principles

Principle 1: role of AIP. AIP should continue to be used in combination with other spectrum management tools, in both the commercial and the public sectors, with the objective of securing optimal use of the radio spectrum in the long term. AIP's role in securing optimal use is in providing long-term signals of the opportunity cost of spectrum.

Principle 2: users can only respond in the long term. The purpose of AIP is to secure the optimal use of spectrum in the long term, so as to allow users to be able to respond to AIP as part of their normal investment cycle. Even where users have constraints imposed on their use of spectrum, in general, some if not all users have some ability to respond to AIP.

Principle 3: when AIP should be applied. AIP should apply to spectrum that is expected to be in excess demand from existing and/or feasible alternative uses, in future, if cost-based fees were applied. In determining feasible alternative uses, we will consider, over the relevant time frame, any national or international regulatory constraints, the existence of equipment standards, and the availability and cost of equipment, as well as other factors that may be appropriate.

Principle 4: the "relevant time frame" to assess future demand of spectrum. In general, we need to determine the time period over which we will seek to assess excess demand, congestion, and feasible alternative use. We will do so over a time frame that reflects the typical economic lifetime of existing users' radio equipment.

Principle 5: AIP and spectrum trading. Many secondary markets are unlikely to be sufficiently effective to promote the optimal use of the spectrum without the additional signal from AIP. Therefore AIP will likely continue to be needed to play a role complementary to spectrum trading for most licence sectors.

Principle 6: role of AIP in securing wider social value. Uses of spectrum that deliver wider social value do not, as a general rule, justify AIP fee concessions, because direct subsidies and/or regulatory tools other than AIP are normally more likely to be efficient and effective.

Principle 7: AIP concessions and the promotion of innovation. It will generally not be appropriate to provide AIP concessions in order to promote innovation.

Principle 8: use of market valuations. We will take account of observed market valuations from auctions and trading alongside other evidence where available when setting reference rates and AIP fee levels. However, such market valuations will be interpreted with care and not applied mechanically to set reference rates and AIP fees.

Principle 9: setting AIP fees to take account of uncertainty. Where there is uncertainty in our estimate of opportunity cost, for example arising from

> uncertainty in the likelihood of demand for feasible alternative uses appearing, we will consider the risks from setting fees too high, or too low, in light of the specific circumstances. When spectrum is tradable we will consider the extent to which trading is expected to promote optimal use, and will also have particular regard to the risk of undermining the development of secondary markets.

In addition, having pre-announced in 2007 its intention to charge scarcity administrative fees for terrestrial broadcasters, including the noncommercial BBC, it withdrew from that proposition in 2013. Introduction of charges was deferred until about 2020 [11].

Such opposition on the part of those on whom spectrum prices are imposed is pervasive, and has in the UK been fairly successful, especially in the private sector. It is a reason in favor of the use of markets. Once a spectrum market has been set up and is in operation, it becomes much more difficult for spectrum users to get round it, as they can sometimes get round administrative prices.

The Australian spectrum regulator ACMA announced a policy of making use of opportunity-cost prices for spectrum in 2010. ACMA expressed a general preference for market methods – auctions and trading, but noted that there was a role for opportunity-cost pricing, especially "where trading volumes are low because of characteristics of the market (e.g. high transaction costs, thin markets, or uncertainty); or if it is useful to ensure that licensees face an actual cost (e.g. for government users who may face lower incentives to trade spectrum)" [12]. ACMA's initial focus was on the 400 MHz band, which is mainly used for the land mobile service, but it also accommodates other services, including the fixed (point-to-point and point-to-multipoint), radiolocation, and amateur services, and is congested in the major cities. However, progress is slow, and some assignments appear to have been made without reference to opportunity-cost pricing [13]. This may reflect the complexity of opportunity-cost pricing in a band with so many multiple uses – the "general-equilibrium" feature referred to above.

To summarize, there are a number of issues which arise in pricing spectrum based on opportunity cost. The most obvious one is the difficulty of identifying and costing alternatives to the band in current use. Then there are complexities, which may not always arise, associated with the multiplicity of ways of using certain bands.

It is sensible to manage the possibility of asymmetric errors in setting administrative prices via an appropriate risk analysis of the kind suggested in Box 7-2, but it is important to remember, when considering the use of opportunity-cost prices, that the relevant alternative to an imperfect administrative price may be a price of zero, which is probably highly inefficient.

7.6 Other pricing applications in practice

As described above, the other form of administered scarcity price sought for spectrum is a proxy for market price, and includes not only scarcity rents captured by the opportunity-cost method but also any monopoly rents and option values, as well as (in principle) the effects of whatever notional auction or market arrangements are assumed. This approach can be used to set a price which spectrum users can be charged. It can also be used to set a reserve price in an auction if the spectrum authority is concerned that otherwise the license in question may not realize its true value as a result of lack of competitors for it or collusion among them (see Chapter 5).

Two methods are considered here for setting administrative prices of this kind – designed to mimic market prices. The first relies on the regulator seeking to reproduce the calculations which might be done by a firm bidding for spectrum. The second relies on drawing inferences from observed market transactions. In principle these could be trading data, but these are so scarce that auction data are generally used.

7.6.1 The business model approach

This works through an attempt by the price setter to replicate the thought and calculation processes of firms seeking to estimate the maximum they can bid for a spectrum license without falling into loss. They are assumed to do this by calculating the net present value of operating the radio system over its life, including a terminal value if appropriate, but excluding the cost of acquiring the license. This calculated value is assumed to represent the maximum bid the firm in question will rationally make.

Clearly this depends crucially both on the overall success of the activity which the spectrum is assumed to support, and on the competitive structure in which the relevant services are supplied, and it will vary from supplier to supplier. The derivation of expected (rather than maximum) bids depends upon the auction rules, the goals of the bidders, and the number of licenses available.[11] But bidders clearly have an interest in avoiding paying right up to their maximum willingness to pay. This step of getting from maximum willingness to pay to amount actually paid is obviously not required when auction revenues can be directly observed from market transactions.

Studies of this kind are challenging. The regulator and its consultants are unlikely to have the detailed sectoral knowledge which a firm in the sector would have; nor, since it is not their money which is being wagered in the auction, do they have incentives for accuracy as sharp as those of bidders. Moreover, the final stage of translating data on distribution of maximum willingness to pay into likely auction outturns is particularly complex.

---

[11] Thus in an auction in which four identical licenses are available, in certain conditions one might expect the outturn uniform auction price to be in the neighborhood of the maximum price payable by the fifth-highest bidder, because that neighborhood contains the price which the fourth-highest bidder has to beat to secure the license.

In cases where spectrum is likely to be used to provide a new service which is already in place elsewhere, it is possible to use analogies with other countries to populate the model of diffusion or even analogous competitive outcomes in the country in question. Thus the transition from analogue to digital terrestrial television has occurred and is occurring in many countries. European experience of this transition permitted one of the present authors rather gingerly to project the process of viewers switching from analogue to digital viewing of existing channels and simultaneously switching some of their viewing to new digital channels, for application to a middle-income country and thus estimate the value of licenses to bidders.

As with the opportunity-cost method described above, after a "best estimate" of the bid outturn has been reached, a decision has to be made as to how to determine the price. This should be done bearing in mind the risk that an overestimate carries the danger of taking the spectrum out of use entirely, as a result of it not being bought at auction, or because the firm to which it is assigned prefers to turn it in rather than to use it.

The business model is a feasible approach, but it requires many assumptions and is accordingly subject to error. If regulators are making several estimates by approaching the figure sought from several angles ("triangulating" it), it may have a place.

7.6.2 Estimates based on auction data

This method works on the basis of inferences which the regulator draws from observed transactions. The goal is to set a price for a spectrum band in one country by analogy with the observed outcomes of past auctions for spectrum either in that country or in other countries, and either in the same band or in other bands. Clearly the inference is likely to be safest when data from a recent auction in the same band in the same country are available and least safe when data are drawn from auctions of dissimilar bands held long ago in remote countries.

We now describe the steps likely to be required in making such an estimate.

7.6.2.1 Standardize the data from different countries

This process relies upon an observation made in work by Hazlett [14], showing that auction revenues, standardized across countries, appear to vary systematically with certain key economic variables. The revenues are standardized by expressing them in a single chosen currency in terms of "per megahertz per population" – by dividing total revenues by the megahertz available under the license and then by the population (pop.) covered in the license. The chief variable used to adjust this value is GDP per head. This is done by establishing the statistical impact of different levels of GDP per head on the sample of observations of revenues per megahertz per pop. Once this relationship has been estimated it is possible to adjust revenues gained in another country to what "they would have been" in the country in which the prices are being set.

7.6.2.2 Standardize across different bands

Spectrum below 1 GHz is worth more than spectrum above it. In order to make use, in setting prices, of auction data relating to different bands, it is possible to establish the relativities observed across bands in countries where auctions in both have occurred. These relativities can be averaged and applied in the country in question.

7.6.2.3 Standardize for license length

In auctions we observe typically a single payment by the winning bidder for licenses of varying lengths. There are various ways of making this adjustment. If the annual benefit derived from the license were uniform over the period, it would be relatively simple to standardize. But this is unlikely, so an alternative procedure may be required. One suggestion is to use a combined approach involving use of auction data and data from a business model.[12]

7.6.2.4 Express the result in annual payments

Spectrum licenses are usually bought at auction in a single up-front payment.[13] Administrative prices are paid annually. Turning a single payment into an equivalent stream of annual payments requires decisions on whether to set a uniform or a varying payment in each year (in nominal terms, or in real terms – allowing for inflation), and what discount rate to use. It is probably simplest to adopt a uniform real annual payment. However, choice of the discount rate is more problematic. One candidate is the firm's cost of capital – its average borrowing rate for debt and equity. However, if the annual charge were a fixed payment, it would be more akin to interest on debt, and should pay a lower rate. This is significant, because a lower discount rate attaches more value to receipts later in the license period, and keeps the annual payments down.

Two examples of this procedure are given. The first relates to the need to set a reserve price for spectrum in an auction in Ireland, conducted by the Irish regulator, the Commission for Communications Regulation (ComReg), and including mobile spectrum in the 800 MHz, 900 MHz, and 1,800 MHz bands. The regulator did not expect additional bidders beyond the existing mobile operators; in addition, caps were imposed on the holdings which operators could acquire. In these circumstances, a reserve price was considered appropriate.

Consultants employed by ComReg used data from previous auctions in broadly the way set out above, in a series of benchmarking studies which took account of new auction data as they emerged [16]. The process as a whole relied heavily on judgment. On this basis the consultants recommended that a range of "conservative" minimum prices be set for the bands – a common range for 800 and 900 MHz and a lower range for 1800 MHz. The regulator chose prices from somewhere near the middle of this range. In the event, all lots were sold to existing operators, at prices close to the reserve prices [17].

---

[12] Suppose the license in country A lasts for 20 years and that in country B for 25 years. To infer a value of spectrum in B from values in A, we standardize for country as described above. The value of the last five years of the license in A is then estimated using the business-model approach (by projecting the operator's earnings in those years). Since these remote revenues have to be discounted, in order to express them in terms of their value 20 to 25 years previously when the auction payment has to be made, their impact on the total valuation is comparatively small. See [15].

[13] This deters an operator from overbidding, and then walking away from the licence when things turn out badly.

In the second case to be considered, the UK regulator was charged with calculating market prices for certain spectrum bands, which would be set as administrative prices. Previously, mobile operators had been charged prices based on opportunity costs for 900 MHz and 1800 MHz spectrum, but in 2010 the UK government directed Ofcom to revise the level of charges to reflect full market value. In doing so, they were to have particular regard to the sums bid in the UK 4G auction, which included 800 MHz spectrum and was to be accomplished in 2013.

As in the Irish case, Ofcom used a variant of the procedure described above in order to "read across" to the problem in hand auction results in other bands or in other countries. And, as in Ireland, the proposed prices were successively reworked as new auction results were announced and further amendments to the method of calculation were made. These proposals, due to come into effect in 2015, would increase charges three- or fourfold, so were not popular with the companies [18].

7.6.3 Summary

The two methods described above provide a practicable means of estimating the market price of a band. Both are clearly fallible. But they can be used in parallel to generate estimates which together may "triangulate" the number sought, and can also be compared with the opportunity-cost method.

7.7 Administrative prices and trading

Administrative prices are usually applied to spectrum, which has not been auctioned, as a means of creating an incentive for its efficient use. Indeed, only in very unusual circumstances would it be good policy unexpectedly to impose *ex post* an administrative price on a license competitively won at an auction. To do so might endanger bids in any later auction.

However, administrative prices can advantageously be combined with spectrum trading for unauctioned spectrum. This has the effect of changing the incentives on the spectrum user to recycle the license. There are three cases:

1. with administrative pricing alone, the incentive to give up spectrum is the price alone;
2. with trading only, the incentive is the cost of holding on to the spectrum, which is the trading revenues foregone;
3. with both at once, the incentive is a combination of price paid and foregone revenue (though that expected trading revenue is, of course, reduced if holding the spectrum entails paying an annual charge).

As always, there is the risk of too high an administrative price, which will take the spectrum wholly out of use. This tends to limit the incentive effect of (1). As between (2) and (3), it is distinctly possible that paying an annual charge has a greater incentive effect than foregoing theoretical trading revenue; if so, (3) may be the best policy.

7.8 Conclusion

This chapter has shown that it is possible in practice to set administrative prices for spectrum, based on either an "opportunity-cost" or a business-model approach. Such prices have the potential to sharpen the incentives for efficiency in spectrum use, provided that the incentives bear down effectively on the user. Chapter 12 discusses the issues this raises in the public sector in particular.

However, our account has identified some fragility in the main ways in which administered prices are calculated. A first consequence of this is that prices should be constructed in ways which take account of the adverse effects of their being too high. A second is that, as in many contexts, an administrative price set by a regulator or equivalent is likely to be inferior, as a price signal supporting an efficient allocation of resources, to a price emerging from a well-functioning market. Administrative prices can also successfully be resisted by spectrum users. These considerations suggest that administrative prices should be used sparingly.

However, in many cases the options are to use either a fallibly estimated administrative price, or in effect to set an "administrative" price of zero – which in the case of valuable spectrum is not a happy outcome in a management regime offering any freedom of action to spectrum users.

References

[1] Report ITU-R 2012, "Economic Aspects of Spectrum Management," Chapter 5.2, at www.itu.int/dms_pub/itu-r/opb/rep/R-REP-SM.2012-1997-PDF-E.pdf.
[2] See http://stakeholders.ofcom.org.uk/consultations/900-1800-mhz-fees/?utm_source=updates&utm_medium=email&utm_campaign=alf-consultation.
[3] "Estimating the Commercial Trading Value of Spectrum: A Report for Ofcom" (2009), 85–87.
[4] M. Cave, "Review of Radio Spectrum Management" (report for the Department of Trade and Industry, 2001), 121–125.
[5] M. Cave, "Independent Audit of Spectrum Holdings (report for Her Majesty's Treasury, 2006), 27–36.
[6] See Table 6.1 at http://stakeholders.ofcom.org.uk/binaries/consultations/srsp/summary/srsp_condoc.pdf.
[7] See Ofcom, "Policy Evaluation Report: AIP" (2009), 29–35, at http://stakeholders.ofcom.org.uk/binaries/research/spectrum-research/evaluation_report_AIP.pdf.
[8] Ofcom, "Decision to Make the Wireless Telegraphy (Licence Charges) (Amendment) Regulations 2013" (May 2013).
[9] Ofcom, "Decision to Make the Wireless Telegraphy (Licence Charges) (Amendment) Regulations 2013" (May 2013).
[10] Ofcom, "Applying Spectrum Pricing to the Maritime Sector, and New Arrangements for the Management of Spectrum Used with Radar and Aeronautical Navigation Aids," Statement (2010), at http://stakeholders.ofcom.org.uk/consultations/aip_maritime/statement.

[11] Ofcom, "Spectrum Pricing for Terrestrial Broadcasting" (2013), statement, at http://stakeholders.ofcom.org.uk/binaries/consultations/aip13/statement/statement.pdf.
[12] See p. 2 of www.acma.gov.au/~/media/Spectrum%20Transformation%20and%20Government/Report/pdf/ACMA%20Response%20to%20Submissions%20Opportunity%20Cost%20Pricing%20of%20Spectrum%20Public%20Consultation%20on%20Administrative%20Pricing%20for%20Spectrum%20Based%20on%20Opportunity%20Cost.PDF.
[13] See www.acma.gov.au/Industry/Spectrum/Spectrum-projects/400-MHz-band/latest-developments-1.
[14] T. Hazlett, "Property Rights and Wireless License Values" (2008) 51 *Journal of Law and Economics* 563.
[15] J. Poort and M. Kerste, "Setting Licence Fees for Renewing Telecommunication Spectrum Based on an Auction" (2014) 38 *Telecommunications Policy* 1085.
[16] DotEcon, "Award of 800 MHz, 900 MHz and 1800 MHz: Fifth Benchmarking Report," a report for ComReg (March 2012).
[17] ComReg, "Results of the Multi-band Spectrum Auction: Information Notice" (November 2012).
[18] Ofcom, "Annual Licence Fees for 900 MHz and 1800 MHz Spectrum: Consultation" (October 2013); Ofcom, "Annual Licence Fees for 900 MHz and 1800 MHz Spectrum: Further Consultation" (August 2014).

Part III

Sharing and other emerging approaches to spectrum management

8 Spectrum sharing and the commons

8.1 Basic approach to commons

8.1.1 Defining spectrum commons

Spectrum bands that do not need a license are said to be "unlicensed" or "license-exempt." Another term for this is "spectrum commons," drawing a parallel with common land, where all are allowed (historically this was used to graze animals). In these bands anyone can transmit as long as they obey a few rules of access, such as maximum power levels. The most popular commons band is at 2.4 GHz, which is used for Wi-Fi, Bluetooth and other similar applications.

8.1.2 Where an unlicensed approach may work

Spectrum regulators have noted that they should not license users where there is no concern about interference. This would be the case where there is more capacity in the radio spectrum than demand and where there are mechanisms for users to self-co-ordinate in some manner to avoid interfering with each other. For many applications there are significant concerns about interference, but there are some special cases. In particular these include:

- Low-power transmissions where signals do not extend very far and so have limited chances of interference.
- Very high frequencies (e.g. above about 40 GHz) where there are large amounts of bandwidth and limited demand because of poor propagation at these frequencies.
- Constrained environments such as indoor operation where the exterior of the building can provide some shielding such that one user does not interfere with another.
- In fixed links where the radio signals are generated as thin beams which are much less likely to interfere with each other than omni-directional transmissions.
- In areas where technology can perform the co-ordination and arbitration of usage. These might be expected to grow over time as technology advances, and various approaches are discussed in this and subsequent chapters.

To date, unlicensed approaches tend to be a mix of both higher frequencies (mostly above 2 GHz) and lower powers (mostly below 100 mW), resulting in low range. This has led to hundreds of different applications and technologies including Wi-Fi, Bluetooth, cordless phones, garage door openers, baby monitors, and much more.

8.1.3 Basic approach to the commons

Regulation of commons, or unlicensed spectrum, is rather like regulation of a park. In a park anyone is allowed in as long as they behave themselves. This means keeping noise down, not annoying others, not having fires or parties, and so on. In some cases these rules are posted at the entrance to the park, while in others they are enshrined in general law.

In the case of spectrum the equivalent rules include low transmit power levels, controlled adjacent-band emissions, and in some cases limited duty cycles.[1] Often these are termed "politeness protocols" [1] by analogy to being polite to others in a public park.

How many parks a city should have is very difficult to determine evidentially, and similarly the right number and size of unlicensed bands is also hard to determine.

8.1.4 Current unlicensed bands

Unlicensed bands to date have currently happened more by accident than by design. The most widely used band is the 2.4 GHz "industrial, scientific, and medical" (ISM) band. This is the frequency at which water molecules resonate and so is most useful for microwave ovens and similar industrial equipment. Because these devices leak some radio waves it was initially thought that this band could not be used for radio communications, and so it was set aside for non-radio "industrial" use. Since the resonant frequency of water is the same around the world, the same band was set aside on a global basis. It became used for research purposes and for deployment of trial systems in academia and similar, where researchers noted that in general the interference generated by industrial equipment could be accommodated with good system design. This led to Wi-Fi proponents and others noting the benefits of use of a globally harmonized unlicensed band of spectrum. More recently the 5 GHz band has been proposed as providing additional Wi-Fi capacity and is generally available around the world, albeit with somewhat different restrictions of use in particular countries.

Other bands have become unlicensed because they were guard bands between licensed use, because they did not "fit" into any appropriate pairing arrangement or similar. Almost all unlicensed spectrum is above 2 GHz (the bands below 2 GHz tend to be very narrow), although notably there is unlicensed spectrum at 900 MHz in the US which is widely used, and a few small bands at frequencies such as 868 MHz in Europe.

Table 8-1 shows the currently unlicensed bands in the UK – other countries, especially in Europe, will have similar allocations.

---

[1] The duty cycle is the percentage of time a device is actually transmitting for. For example, a device that transmitted for one second in every 10 seconds would have a duty cycle of 10%. In unlicensed bands duty-cycle restrictions are typically in the range of 1%–10%.

Table 8-1. Unlicensed bands in the UK

Generic frequency band	Application
9 kHz–30 MHz	Short-range inductive applications
27 MHz	Telemetry, telecommand, and model control
40 MHz	Telemetry, telecommand, and model control
49 MHz	General-purpose low-power devices
173 MHz	Alarms, telemetry, telecommand, and medical applications
405 MHz	Ultra-low-power medical implant devices
418 MHz	General-purpose telemetry and telecommand applications
458 MHz	Alarms, telemetry, telecommand, and medical applications
864 MHz	Cordless audio applications
868 MHz	Alarms, telemetry, and telecommand applications
2,400 MHz	General-purpose short-range applications, including CCTV and RFID. Also used for WLANs, including Bluetooth applications
5.8 GHz	HiperLANs, general-purpose short-range applications, including road traffic and transport telematics
10.5 GHz	Movement detection
24 GHz	Movement detection
63 GHz	2nd phase road traffic and transport telematics
76 GHz	Vehicle radar systems

Source: Spectrum Strategy Document, published by the Radiocommunications Agency.

8.1.5 Could everything be commons?

Some have noted that a relatively small amount of spectrum at 2.4 GHz (a bandwidth of 80 MHz) can carry a vast amount of data [2]. As a result it appears to be highly efficient usage of the spectrum when measured in terms of metrics such as bits per second per hertz per kilometer squared. This is generally more technically efficient than cellular or broadcast systems, leading some to argue that if all spectrum were unlicensed the efficiency of use would be much greater and there would no longer be a shortage of spectrum (and hence a need to license it).

While Wi-Fi is technically very efficient, it would not be an economically efficient alternative to cellular radio. The number of Wi-Fi hotspots needed to provide equivalent coverage would be impractically high. Instead, systems with greater range are needed, but these then have a higher likelihood of interference. Hence they need to be licensed. So the short answer is that, at least today, not everything can be unlicensed. It is possible that technology may evolve sufficiently in the future that the extent and scope of licensing could be relaxed somewhat, but this will likely happen over many decades.

8.2 The tragedy of the commons

8.2.1 The fears

Analogies are often drawn between spectrum and land. The equivalent of open-access spectrum in land today is parkland, but historically it was land that could be used for grazing by all.

In English villages (as with mountain countries in Europe), shepherds sometimes grazed their sheep in common areas, and sheep ate grass more closely than cows. Overgrazing could result because for each additional sheep, a herder could receive benefits, while the group shared damage to the commons. If all herders made this individually rational economic decision, the commons could be depleted or even destroyed, to the detriment of all. This is known as the "tragedy of the commons," and we see it happening in many areas where individually rational decisions are to the detriment of society as a whole – runs on banks, for example, fall into this category.

So by analogy many expect commons or unlicensed spectrum to be overused by individuals to the point where it is useless to all because the levels of interference are so high that no useful communication can take place.

However, this has not happened to date. There are a number of reasons for this, which broadly relate to ways that the technology itself is designed in such a manner as to reflect the common good rather than individual needs. These include the following:

- Technologies tend to be "listen-before-talk," so do not tend to consume resources when others are transmitting.
- Where there is interference, e.g. at one point between Bluetooth and Wi-Fi, the standards bodies tend to work together to resolve the issue and update the technology appropriately. In this case Bluetooth detects Wi-Fi and rearranges its hopping pattern to avoid it [3].
- Individuals can co-ordinate their own multiple uses of the band; for example, where they buy products that do interfere with each other, they tend to send them back.
- The limited range associated with the rules of usage helps keep interference a personal rather than a shared matter, avoiding the problems that led to the tragedy of the common land.

The reason why those designing the technology put the "crowd" before the "individual" and adopted solutions that share well was not that they were worried about social responsibilities, but more that they wanted their technology to succeed in widespread usage and could see as a result that it would need to work in such a way as equitably to divide up any resource in the case of congestion.

Hence, while many still fear that the success of an unlicensed band will lead to it becoming too congested to be of any use, there are good grounds for believing that this will not happen.

It is instructive to compare spectrum commons with work by Elinor Ostrom which identified eight "design principles" of stable local common-pool resource management [4]:

1. clearly defined boundaries (effective exclusion of external unentitled parties);
2. rules regarding the appropriation and provision of common resources that are adapted to local conditions;
3. collective-choice arrangements that allow most resource appropriators to participate in the decision-making process;
4. effective monitoring by monitors who are part of or accountable to the appropriators;
5. a scale of graduated sanctions for resource appropriators who violate community rules;
6. mechanisms of conflict resolution that are cheap and easy of access;
7. self-determination of the community recognized by higher-level authorities; and
8. in the case of larger common-pool resources, organization in the form of multiple layers of nested enterprises, with small local CPRs at the base level.

These principles have since been slightly modified and expanded to include a number of additional variables believed to affect the success of self-organized governance systems, including effective communication, internal trust and reciprocity, and the nature of the resource system as a whole [5].

We consider the extent to which these rules have been met in the spectrum commons in Table 8-2.

Interestingly, the rules are partially met, but the key difference is that many of the functions of monitoring, self-determination and organization are performed by the communications technology, not by individuals, and the rules for operation can be codified. This suggests that sound regulatory decisions from standards bodies and spectrum managers can broadly avoid the tragedy of the commons.

8.2.2 The reality to date

Actually understanding whether there is congestion in unlicensed bands is very difficult. There is no easy way to measure usage and the impact of congestion and it can vary dramatically, for example from high levels within an airport terminal to very low levels in the airport parking lot. It can also vary dramatically by time of day. Where systems such as Wi-Fi are working slowly the cause may not always be apparent and might be to do with aspects unrelated to spectrum.

Most measurements to date [6] suggest that interference or congestion in the 2.4 GHz band is very rare and is restricted to high-density public areas such as shopping malls and stadiums. Even here, problems can generally be overcome with careful deployment of an ever-higher density of base stations. So despite the proliferation of Wi-Fi and Bluetooth devices, a tragedy of the commons has not generally occurred (although there have been a few specific situations where problems have arisen [7]). Congestion in licensed spectrum is just as likely as congestion in unlicensed.

However, it cannot be assumed that a problem will never occur. The use of lower frequencies with resulting longer range will increase the propensity to interfere and make the problem more of a collective than an individual one. Ever more unlicensed

Table 8-2. Comparison between Ostrom's principles and unlicensed spectrum usage

Ostrom principle	Spectrum usage
Clearly defined boundaries (effective exclusion of external unentitled parties)	Yes – by defining the technologies allowed to use the spectrum
Rules regarding the appropriation and provision of common resources that are adapted to local conditions	Generally yes – rules are band-specific and may differ from country to country
Collective-choice arrangements that allow most resource appropriators to participate in the decision-making process	Partially – the license conditions and standards are defined by a few, but they do have the interests of the many at heart when making their decisions
Effective monitoring by monitors who are part of or accountable to the appropriators	Partially – the technology itself tends to monitor (e.g. listen before transmitting) and usage parameters can be derived by regulators
A scale of graduated sanctions for resource appropriators who violate community rules	Partially – enforcement proceedings can be made against those operating outside the license terms
Mechanisms of conflict resolution that are cheap and easy of access	Partially – these are embedded within the technology itself
Self-determination of the community recognized by higher-level authorities	Partially – self-determination is performed by devices sanctioned by standards bodies and regulators
In the case of larger common-pool resources, organization in the form of multiple layers of nested enterprises, with small local CPRs at the base level	Partially – spectrum use within buildings, for example, is smaller in scale and local

devices in the home might cause increasing issues, especially if homes need multiple different networks (e.g. one for home automation, another for entertainment, etc.). Regulators need to be constantly aware of the possibility of interference and to monitor and adjust usage as far as possible. This might include making more spectrum available in other bands or looking to adjust the rules of entry for future devices.

8.3 Restriction on usage in various bands

8.3.1 Ways to restrict usage

All unlicensed bands have "rules of entry." These are designed to restrict usage in some manner in order to prevent interference occurring.

The simplest restriction is reduced power levels. By lowering the power the range of transmissions is decreased, hence the zone over which interference can occur shrinks. This means fewer people are affected by any one transmission and the overall capacity of the band is improved, albeit at the expense of lower utility for each individual – for example, if this now means that a Wi-Fi router is unable to provide coverage throughout a house this could be annoying for the homeowner.

Another restriction is to limit what is known as the duty cycle. This is the percentage of time that a device is transmitting. It would normally be measured over an interval such as 10 seconds, and duty cycles are sometimes as low as 1% – meaning that a device can only transmit for 1% of the time every 10 seconds. This prevents one device from monopolizing the spectrum in any area. It also reduces the overall usage of the spectrum by rendering some applications impossible. For example, Wi-Fi routers cannot work with very low duty cycles as they need to periodically broadcast beacon information. Duty-cycle restrictions are often used in narrow frequency bands where just one or two devices could otherwise easily use the entire band.

A third restriction is to limit the maximum bandwidth allowed. This prevents one device from using the entire spectrum with a single transmission. The combination of power, duty-cycle, and bandwidth restrictions limits the amount of information that can be transmitted by any application.

Finally, various politeness protocols might be imposed. Most popular is listen-before-talk, where devices have to monitor usage of the band and can only transmit if the band appears to be free. In some cases this has been introduced to avoid interference with prior licensed use of the band, but it can also ensure that interference levels remain low within the band.

In principle, all of these restrictions could be dynamic, changing over time based on the congestion levels currently being experienced, although that would require signaling mechanisms and standardized approaches.

8.3.2 Light licensing approaches

Another type of restriction is to require some form of registration [8, 9]. This typically goes under the title of "light licensing" and is generally the provision of access to all who want it (often free of charge). However, some record is kept of deployments to allow co-ordination between users. This can improve interference mitigation and can provide some increased certainty of quality of service. Registration of usage also allows the regulator to understand the level of activity in the band, which can be helpful management information in determining whether the band is becoming full and whether any spectrum management actions need to be taken.

Registration may also be required by the licensed user where the unlicensed user is sharing their band. For example, some light licensed use is in bands owned by the military and the provision of registration makes it easier for them to address any issues which might occur.

The ECC have defined light licensing thus:

A "light licensing regime" is a combination of license-exempt use and protection of users of spectrum. This model has a "first come first served" feature where the user notifies the regulator with the position and characteristics of the stations. The database of installed stations containing appropriate technical parameters (location, frequency, power, antenna etc.) is publicly available and should thus be consulted before installing new stations. If the transmitter can be installed without affecting stations already registered (i.e. not exceeding a pre-defined interference criteria), the new station can be recorded in the database. A mechanism remains necessary to enable a new entrant to challenge whether a station already recorded is really used or not. New entrants should be able to find an agreement with existing users in case interference criteria are exceeded. [10]

The need to record the location of devices in a database tends to bias light-licensing applications towards fixed usage, including

- fixed links,
- satellite systems, and
- networks with base stations.

It tends to be less useful for consumer applications as consumers will generally not want to register or co-ordinate.

Light licensing is used where licensing is not needed because interference is unlikely but no interference can be tolerated in any case. Fixed links are a classic case where each link can be licensed individually and interference between them is unlikely, but many users feel the need for some protection against the unlikely case of another link being in the same place and frequency.

Because database entries can be refused in particular geographical areas it can also be a good way to work around licensed users with specific deployments (e.g. satellite uplink sites). However, it does require an element of trust that the system is deployed where stated and is not moved.

The simplest approach to light licensing is to require users to register in a database open to all. (Access can be restricted to pre-checked users or similar where there are confidentiality concerns.) Existing users hope that new users check the database and refrain from a deployment that will interfere with an existing use. There is an element of self-interest in this since interference is normally mutual, so a user deploying an interfering link will likely suffer interference themselves.

A more complex database could perform a check for any new registration and advise if there would be interference. This might be helpful since many users will have limited ability to conduct such checks for themselves, but it does require definition of propagation models and understanding of interference scenarios, and if interference still occurs may lead to some ambiguity as to who is responsible.

An even more interventionist approach is to put in place regulation such that if a new entrant causes interference they must resolve it. This would often mean the new entrant turning off their transmitter or moving it in space or frequency. The onus is on the existing user to spot the interference and notify the regulator, at which point the registration date for the system or similar establishes who was there first (this can fail where systems are registered long before deployment). However, this is starting to get into difficult areas such as defining what constitutes interference – although the threat of having to resolve interference is often enough to ensure that it is not caused in the first place.

Another option is to require all those using the band to agree among themselves on the rules for usage and means of mitigating interference. If they are unable to agree then the regulator might impose conditions (this threat is typically sufficient to cause the band users to reach agreement themselves). This can only work with a relatively limited number of players (perhaps ten or less), otherwise the negotiations tend to be very hard. It also generally requires most or all users to be identified at the start of the use of the band as any subsequent users do not get to contribute to the development of the rules. This approach was adopted for the so-called "DECT guard band" in the UK [11] – a small piece of spectrum that had previously been set aside as a guard band between DECT and GSM1800. Subsequent study determined that it was unnecessary and could be used for low-power GSM transmission, and it was auctioned to around eight users who could all access the entire band, sharing it among themselves. However, having paid their auction fee, all but one player decided not to proceed and so the rules for co-operation were never tested.

A further evolution would be for a device to determine its location, query the light-licensing database, get back information on whether there would be interference for the proposed usage, then go ahead and make the deployment. This kind of dynamic spectrum access using a database is precisely what is envisaged for "white space" access described in the next chapter, and indeed it might be expected that dynamic access will gradually replace light licensing, eventually making it redundant.

So in summary, light licensing plays a role where congestion is unlikely but users require certainty that, once installed, a system will not suffer interference or a drop in capacity – fixed links often fall into this camp. The category covers a broad range of different approaches, some of which may continue to be useful, but it is likely to be increasingly overtaken by dynamic spectrum-access approaches such as white space.

8.4 The Ofcom Licence-Exemption Framework Review

In 2007 Ofcom looked into a framework for unlicensed use [12]. They considered a number of general points around the rules for managing unlicensed bands and the cases where access should be unlicensed. Their recommendations fell into four areas, as discussed in the following subsections.

8.4.1 Application specific?

The first question considered was the extent to which unlicensed bands should be used for specific applications – for example, whether the 2.4 GHz band should be dedicated to Wi-Fi usage.

Application-specific spectrum reduces the probability of interference in two ways:

- Interference between devices using the same application is typically much less problematic than between different applications because the technology or standard will normally have been designed with the assumption that interference from other like devices will occur.
- By restricting the application, fewer devices might be expected (e.g., in the case of 2.4 GHz, if it were dedicated to Wi-Fi then Bluetooth devices would be excluded).

However, there are many downsides. There is less flexibility in the use of the spectrum and if the application does not succeed then the spectrum is left unused. Even if successful the application might not be the best use of the spectrum. There is a risk of multiple bands being needed, one for each application, which would be difficult to find and manage. Generally, the best way to promote innovation and generate value is not to restrict the range of applications and technologies that are allowed to use the spectrum, and allow the market (rather than the regulator) to decide best use.

So there is a difficult balance between prescriptive measures to manage interference, which may be effective but limit innovation, and general measures or no measures, which allow innovation but may be ineffective. To date interference has not been a serious problem in unlicensed spectrum, suggesting that a light-touch approach is better. Hence, Ofcom concluded that bands should not be set aside for specific applications. However, they did note that interference between applications tended to become increasingly problematic the more difference there was between the power and time utilization of spectrum. So, for example, an application like Wi-Fi that transmitted at relatively high power for much of the time would not tend to share well with an application such as short-range radio frequency identification (RFID) tags, which are very low-power and make occasional short transmissions. In this case, the Wi-Fi would tend to drown out the RFID tag. As a result, Ofcom suggested that unlicensed spectrum be divided into different classes, of which there might be a small number (e.g. three or four). Some bands would be for high-power/high-duty-cycle devices, while others would be for medium- and low-power devices. The number and size of each would depend on need and might be expected to change over time.

To date, these proposals do not appear to have been implemented, although there is some division of applications/powers already, in that, for example, there are dedicated bands in Europe for RFID tags.

8.4.2 Politeness rules

Ofcom then suggested that within any band devices should be polite. It broadly defined polite operation as:

- using the lowest possible transmit power via techniques such as transmit power control,
- transmitting for the shortest possible length of time and ceasing transmission when there were no data to send, and
- sensing the usage of a band and not transmitting when others were using the band and when to transmit might be expected to cause them interference.

Ofcom decided that it could not be more specific about politeness. For example, it could not insist on particular forms of transmit power control or particular avoidance mechanisms. To do so would limit innovation and might be inappropriate for future technologies. Even specifying these characteristics for known current applications would be difficult.

Instead, Ofcom felt that the specific politeness approach for any given technology should be defined by the standards body or similar. However, they should consult the regulator, who would decide whether the proposed approach met the intent of the politeness rules or not. This would involve case-by-case judgments about which it would not be possible to provide any specific guidance in advance. Nevertheless, it seemed likely that such an assessment would be relatively straightforward in most cases.

This approach does not appear to have been implemented either, although standards bodies working on technologies for unlicensed bands do tend towards polite approaches.

8.4.3 Higher frequencies

Ofcom noted that at higher frequencies there should be increased propensity towards unlicensed use since there was much less possibility of congestion. They suggested that "higher frequencies" might nominally be considered to start at around 40 GHz. Between around 40 GHz and 100 GHz much of the spectrum was allocated to existing uses, but any that was not should be made available on an unlicensed basis. This included a band from 59–64 GHz. Other bands might have licensing conditions progressively eased over time as appropriate.

Above 100 GHz almost all spectrum should be unlicensed except where there was pre-existing usage.

While this opens up a lot of spectrum by bandwidth, there are currently few applications that wish to make use of this spectrum and hence its economic impact will be slight. However, it does enable possible future innovation.

8.4.4 Lower powers: UWB and Part 15

Finally, Ofcom suggested that some emissions were at such low power that they could be exempt from licensing regardless of the frequency band which they used. The US has for a long time had a regulation of this sort called "Part 15" [13], which allows transmissions below −41 dBm/MHz to be exempt. Originally this approach reflected the fact that many electronics devices were unintentional transmitters – for example, computers tend to emit radio signals at the clock frequency of the key processors used on board despite the best effort of the designers to prevent this. As long as these levels are sufficiently low they are judged acceptable and no radio license is needed.

Such lower power levels were not thought to be useful to communications until a new approach was suggested called ultra-wideband (UWB). This makes use of very broad bandwidth (sometimes as much as 1 GHz) to increase the effective range (recall that the information content for a given range is dependent on the bandwidth and power level, or, conversely, for a given throughput requirement and power level, recall that the range can be increased by increasing the bandwidth). Even so, UWB still only offered a range of between 1 m and 10 m depending on the data rates required. For a range of reasons UWB was not successful commercially, although it is used for some niche applications such as ground-probing radars. However, it is possible that it might find some use in the future.

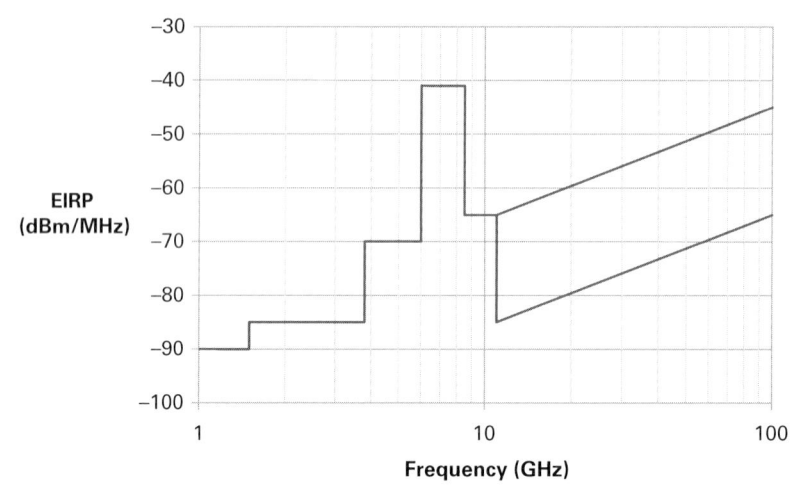

Figure 8-1. Ofcom's proposal for low power exemption.

Ofcom's approach was not as simple as that of the FCC. Instead, they proposed the mask shown in Figure 8-1. This complex-looking mask was based on previous work that Ofcom had completed on UWB, considering the interference potential with each of the applications currently using the bands below 10.6 GHz. Only above 6 GHz was the US level of −41 dBm/MHz allowed. Ofcom then extrapolated powers from their level at 10.6 GHz upwards using a sloping line to reflect the decreased propagation at higher frequencies. They proposed two lines, the lower one to apply to bands where there were passive (receive-only) services – typically radio astronomy – and the upper line to apply in all other cases.

8.5 Summary

Where there is unlikely to be congestion and interference in the use of a band of radio spectrum (so that supply exceeds demand) there is generally no need to license users. Instead, an unlicensed, or "commons," approach can be adopted where the regulator sets the basic rules of access to a band and then allows all to use it. The rules generally aim to ensure "polite" behavior between the users such that the spectrum can be shared equitably.

With any commons there is a risk of a tragedy of the commons where individual self-interest acts against the interest of the group as a whole. This does not appear to have happened in radio spectrum, probably because the technology itself tends to act in the interests of the group more than it does in the interests of an individual. Unlicensed spectrum appears to be generally lightly used despite its ever-increasing utility for applications such as Wi-Fi and Bluetooth. However, regulators do need to monitor this and be prepared to act if congestion should arise.

An alternative to an unlicensed approach is to use registration of users – this is often called "light licensing." This can bring a number of benefits such as allowing geographical exclusion zones, providing increased certainty from interference and allowing users to co-ordinate between themselves. It might be expected that light licensing will gradually move towards dynamic spectrum access.

Ofcom studied the use of unlicensed spectrum in their Licence-Exemption Framework Review. There they concluded that bands should not be application-specific but that like uses (in terms of their transmit powers and duty cycles) should be grouped together. They suggested that polite protocols be adopted by standards bodies. They thought that most spectrum above around 40 GHz should become unlicensed and that there were lower power limits below which licensing was unnecessary.

References

[1] See discussion at http://stakeholders.ofcom.org.uk/market-data-research/other/technology-research/research/exempt/polprot.
[2] See, for example, David Reed's webpage at www.reed.com/dpr/locus/OpenSpectrum.
[3] See www.ee.ucl.ac.uk/lcs/previous/LCS192003/125.pdf.
[4] E. Ostrom, *Governing the Commons: The Evolution of Institutions for Collective Action*, Cambridge University Press, 1990.
[5] J. Poteete and E. Ostrom, *Working Together: Collective Action, the Commons, and Multiple Methods in Practice*, Princeton University Press, 2010.
[6] See Ofcom's pioneering work in this space at http://stakeholders.ofcom.org.uk/market-data-research/other/technology-research/research/exempt/Wi-Fi.
[7] E.g. see www.wired.com/2010/06/wireless-woes-rain-fail-on-steve-jobs-keynote.
[8] ECC, "Light Licensing, Licence-Exempt and Commons," at www.erodocdb.dk/Docs/doc98/official/pdf/ECCRep132.pdf.
[9] E-band Corporation, "Light Licensing," at www.e-band.com/get.php?f.848.
[10] ECC Report 132, at www.erodocdb.dk/Docs/doc98/official/pdf/ECCRep132.pdf, section 2.3.2.
[11] Ofcom, "Proposal for the DECT Guardband," at www.ofcom.org.uk/static/archive/ra/topics/pmc/consult/gsm1800/gsm1800condocfinalweb.pdf.
[12] Ofcom, "Licence-Exemption Framework Review," at http://stakeholders.ofcom.org.uk/binaries/consultations/lefr/summary/lefr.pdf.
[13] See FCC, "Title 47: Telecommunication," Chapter 1, Subchapter A, "General," Part 15, "Radio frequency devices."

9 Dynamic spectrum access

9.1 Introduction

The previous chapter touched on the concept that devices could share spectrum on a "dynamic" basis. So, for example, if a primary user was not using spectrum in a given location at a particular time then a secondary user might opportunistically make use of it. This could be considered to be an extension of light licensing where the database recording usage became more real-time and able to perform coexistence calculations.

The rationale for a more dynamic access is that despite the apparent scarcity of spectrum, observations of actual usage appear to suggest that it is far from full. Observing spectrum usage is a relatively crude process. It involves setting up a receiver that can scan through a range of frequency bands and noting those bands in which energy was received. These are assumed used, whereas the bands where no energy was seen are assumed empty. There are many problems with this approach:

- The receiver may not be able to hear the signal because of local blocking. For example, uplink signals from mobile phones to a nearby base station would be very difficult to receive at a scanning receiver unless it was located close to the base station.
- The band may be used for applications such as radio astronomy where the absence of man-made signals is necessary.
- The band may be used for satellite transmissions and unless the scanning receiver has an upwardly pointing dish it may not receive these.
- Applications such as radar can be very difficult to receive as they only emit short pulses of energy and the probability of the pulse coinciding with the radar pointing at the scanning receiver and the scanning receiver being tuned to the radar frequency at that point in its scan is low.
- Many applications require that the same frequency is not used in neighboring cells. Hence only about 25% of frequencies can be used in one cell, with these frequencies being reused in distant, noncontiguous cells. So although 75% of the spectrum is unused, it is not available to other similar-power applications as that would lead to interference.

So measurements of utilization using signal levels will inevitably underestimate the usage of the spectrum. Multiple measurements across many cities have all tended to show that only around 20% of the spectrum appears to be utilized.[1]

---

[1] For example, see the Ofcom Spectrum Framework Review.

If 20% occupancy were really the case then the availability of a fivefold increase in spectrum usage could be of great assistance in easing spectrum congestion. But the discussion above suggests that we would only expect to see something below 25% utilization because of the need to reuse frequencies coupled with the difficulties of receiving certain types of signal. A measurement of 20% on this basis seems consistent with the spectrum generally being well used. Of course, this is not to say that there is not scope for efficiency improvements, but that these will need to be carefully sought and may not be large.

9.2 Approaches to dynamic access

In order to use spectrum dynamically a device needs to be able to determine whether the spectrum is currently in use in its location – or more specifically whether the device transmitting it will cause interference to the primary or licensed user. Three different ways to do this have been proposed – sensing, beacons, and geo-location. These are discussed below.

9.2.1 Sensing

The approach first suggested was one of sensing. A device that wishes to use the spectrum would scan across a range of frequencies and note those where no activity was detected. It will then assume these to be free for use and transmit accordingly. However, as noted earlier, it can be very difficult to detect usage of a frequency. This is known as the "hidden-terminal problem" [1] because there is a terminal that is hidden to the sensing device often as a result of it being shielded, perhaps by a building. When this hidden terminal transmits, its signals may propagate to the device it is trying to communicate with but not to the sensing device. The sensing device might then erroneously assume the spectrum was free, transmit, and cause interference.

Another way of looking at this is a classic TV reception case, shown in Figure 9-1. Here, a rooftop-mounted antenna receives a weak TV broadcast signal. Because the receiving antenna is elevated above the clutter and has directionality it is able to gather sufficient signal strength to allow the TV signal to be useful. A sensing device might be down amongst the urban clutter and with a nondirectional antenna. The signal it will see will be very much less than the TV receiver and it may conclude that the TV channel is not in use. If it were then to transmit on that channel, its transmission might be very close to the TV receiver and hence little attenuated, causing major interference.

The way to resolve this issue is to make the scanning receiver much more sensitive so it can detect signals even when they are very weak. But in practice this is very difficult to do. In the case of the TV receiver shown in Figure 9-1 the signal level at the rooftop might be some -80 dBm. Extensive measurements and modeling [2] have shown that in 1% of the cases the signal levels that the scanning device receives could be 35 dB or more below this level. This would put the signal level at -115 dBm while the noise floor in the band is about -105 dBm. Detecting signals below the noise floor is very difficult to do. It requires searching for repetitive features in the signal and correlating against these

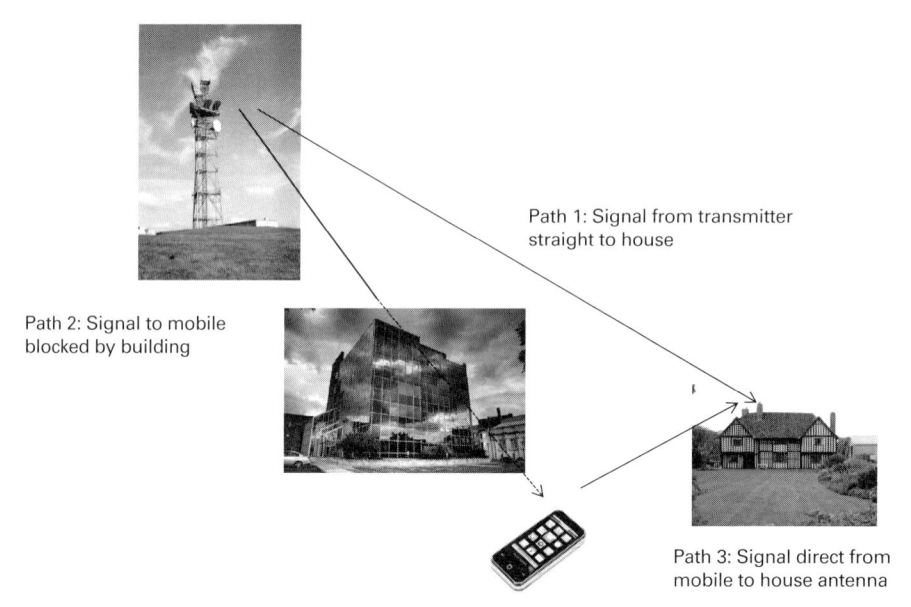

Figure 9-1. The hidden-terminal problem. Source: Ofcom.

in order to bring the signal up above the noise floor. It also needs highly sensitive receivers with extremely good isolation from noise sources generated in other parts of their circuitry. While TV signals do have some repetitive elements, these vary according to the standard and to the particular variant of the standard (e.g. DVB-T versus DVB-T2) and are generally insufficient to allow accurate identification of the signal at the levels needed.

As a result of the hidden-terminal problem it does not appear possible to design and manufacture commercial dynamic spectrum access (DSA) devices that have the sensitivity needed to accurately identify when a frequency is in use. This was a conclusion reached in the US by the FCC [3] and others after a set of equipment trials, and by Ofcom in the UK [4] after modeling and industry consultation. Hence, for many sharing applications, sensing is insufficient.

There are some exceptions. Military users may not be overly concerned about causing interference to hidden terminals as, for example, they move onto a battlefield. In this case sensing may be a viable solution and indeed does appear to be used in military DSA equipment. It is possible that other cases where interference tolerance is relatively high will also appear in future.

9.2.2 Beacons

An alternative to the sensing approach is to use a more failsafe "transmit on receiving a signal that confirms you can do so." DSA devices listen out for these "beacon" transmissions and decode the information within them to determine which channels they can use. This does not rely on the quality of the DSA device to protect the licensed user since a device with an insensitive receiver will not be able to pick up the beacon when further away from it and as a result will have less area over which it is able to transmit.

However, beacons give rise to a number of problems:

- beacons need a spectrum allocation themselves;
- someone needs to build and maintain the network of beacons;
- the beacon transmission may not be accurately constrained to its desired area; and
- protocols need to be agreed at least nationally, if not globally.

The beacon spectrum allocation can be relatively narrow since the beacon will typically not transmit much information, hence the spectrum used could be small. Nevertheless, it may be difficult to find a small slice of bandwidth within a band plan such as that used for DTT, which comprises 6 MHz or 8 MHz channels.

Building and maintaining a network may be costly, although there should be opportunities for sharing the same infrastructure as the service currently using the band since its transmitter towers might be the ideal location for the beacon transmitters. This cost could be problematic if the DSA is unlicensed and therefore difficult to apply any fees to.

Perhaps most problematic is accuracy of indication. DSA devices will transmit wherever they hear the beacon, hence the beacon coverage must be shaped to the areas where the spectrum is unused. But shaping coverage is difficult and receivers with differing levels of sensitivity will be able to pick up the beacon in differing areas. Because there is generally a strong desire to avoid interference with the licensed user, the inclination is to make the beacon's coverage area smaller than that desired. Then, if the beacon does propagate further than expected or better-performing devices are used, the DSA device may still be within the allowed transmission contour. This will mean that, for the average DSA device and the average propagation conditions, the area available for transmission will be smaller than optimal. If DSA coverage areas are relatively small, or oddly shaped, or change frequently, then the beacon approach may result in very inefficient usage of the supposedly spare spectrum.

There was much study of beacons under the term "cognitive pilot channel" in the late 2000s from a range of EU-sponsored research programs,[2] and hence many publications are available. But despite this research, and because of the issues identified above, the concept of beacons has not been progressed further and seems unlikely to form a significant part of the DSA solution.

9.2.3 Geo-location

Another alternative is not to leave the determination up to the DSA device at all. Instead, all determination of free spectrum is done centrally, in a "database" which contains enough knowledge that, given the location of the DSA device and

---

[2] See, for example, the COGEU project [5].

knowledge of its characteristics, it can determine which frequencies are available. This approach is known as a geo-location database. It starts with the device determining its location, using methods such as GPS, or for a fixed device a pre-programmed location. The device then sends this information to the database using "conventional" communications channels such as available backhaul or a cellular data channel. (At this stage it is unable to use dynamic access.) The database then sends back to the device, still over a "conventional" channel, the frequencies that are available in that location. The device decides which frequency to use and then starts to operate on that frequency.

The database needs to have available to it the details of the licensed usage of the spectrum such as the transmitter locations or known coverage areas. It also needs to understand the susceptibility of the licensed use to interference from the DSA device. It is typically provided with a set of rules by the regulator which it then uses to determine how much spectrum is available.

This approach avoids all the problems associated with sensing or the need to transmit a beacon. It also brings some key advantages:

- because the devices are under the control of the database their behavior can be altered by the regulator or other entities should there be any interference problems,
- future changes in spectrum bands that the devices are allowed to access can readily be communicated,
- various forms of reservation or device control could be implemented to prevent excessive interference between DSA devices, and
- band utilization levels can be approximated based on the volume of database requests.

However, there are some disadvantages:

- devices need to be able to locate themselves (or be located) and need an alternative communications path to contact the database,
- a database needs to be set up and run by some entity, and
- it is possible that the records in the database do not reflect well the actual radio transmissions experienced, resulting in problems occurring "on the ground" (although these can be corrected once discovered).

These disadvantages tend to bias the users of geo-location databases towards those entities running networks of base stations since these have a known location and permanent backhaul. The terminal devices in such a network can then operate as slaves so they do not need to geo-locate or contact the database.

Geo-location approaches are the only ones that appear workable at present, at least in those situations where the licensed users must have a high level of protection from interference. Hence, it is the approach currently adopted around the world by those regulators that allow DSA.[3]

---

[3] The Dynamic Spectrum Alliance – at www.dynamicspectrumalliance.org – is a good source of information on regulatory progress around the world.

9.3 Licensed shared access

If dynamic access to the band is possible it can be enabled either on a licensed or an unlicensed basis (in just the same way that access to unused spectrum could be licensed or unlicensed). In this section we consider licensed access on a shared basis, or licensed shared access (LSA).[4] The following section considers unlicensed shared access (often known as "white space" access).

9.3.1 Explaining LSA

The concept of LSA is to allow a limited number of secondary users into a band on a licensed basis. This might be just one other user in some cases. LSA is currently primarily foreseen as a mechanism to enable mobile broadband operators to access spectrum that has been harmonized in their region for mobile broadband use but where there are incumbents that are difficult to "evict." The idea is to award a license, similar to an exclusive license at e.g. 800 MHz, but with the requirement to share with the incumbent. This approach is particularly useful where the incumbent is a governmental user such as the military or aeronautics. Frequency bands under consideration for LSA include the 2.3 GHz band and the 3.4–3.8 GHz band.

LSA would operate in a very similar manner to white space. The regulator would define the access rules in conjunction with the incumbent and would facilitate a geo-location database that would provide the shared user with the rights to access the spectrum in given locations. The difference would be that the regulator would then award the access rights to a limited number of users, perhaps using an auction or similar.

LSA has been proposed by many of the key cellular manufacturers and is currently under consideration at a European level.[5]

9.3.2 Reasons for licensing

The decision of which approach to adopt might also be informed by the nature of the primary licensed usage. For example, if the primary use is intensive then sharing opportunities will be limited. Any significant network rollout would be unlikely, leading to local applications that may be less concerned about certainty of investment. In this case an unlicensed approach may be preferred. Where the licensed use is sensitive, e.g. military usage, then the primary user may prefer a licensed approach where they have close interaction with the users. Where the band is harmonized for mobile broadband but there is an incumbent non-broadband user, then, given that mobile broadband operators tend to prefer licensed access, LSA might be a better approach.

Assigning LSA rights may be problematic. The regulator may have little insight into what the primary user will do in future, such as whether they might significantly expand their network. Hence, the guidance given in any auction would be vague, requiring the bidders to take significant risk.

---

[4] Sometimes known as authorized shared access (ASA).   [5] See, e.g., the RSPG opinion on LSA, [6].

Overall, this suggests a case-by-case approach to deciding whether to use unlicensed or licensed shared access, according to

- the amount of spectrum available for sharing – if limited geographically or temporally, prefer unlicensed access;
- the certainty of future use by the licensed user – if uncertain, prefer unlicensed access;
- the need for the licensed user to tightly control unlicensed use – if high, prefer LSA;
- general views on licensed–unlicensed balance in relevant frequency bands – if little unlicensed spectrum is available, prefer unlicensed access;
- the preferences expressed by most likely secondary users;
- the ability of the licensed user to give out secondary rights directly (e.g. by trading or leasing) – if high, then leave this to the market to handle directly.

This is an area that is still to be studied and tested. It seems likely that, as DSA becomes more popular, and if it moves to other bands aside from TV white space, semi-licensed approaches will increasingly be tried. Happily, the database approach allows for such evolution, with pricing mechanisms being layered on top of conventional database access if and when they are needed.

9.3.3 Current status

At the time of writing in 2015 LSA was still at the proposal stage and had not been implemented. It is unclear whether it is of sufficient interest to be adopted.

9.4 Unlicensed shared access

9.4.1 Explaining unlicensed access

In unlicensed shared access any user can share the spectrum owned by a license holder as long as they do not interfere with the license holder. This is typically ensured using a geo-location database. There are no restrictions on the number of unlicensed users or on their use of the spectrum as long as their equipment can be shown to conform to the rules of access. This is the approach proposed for the TV bands (approximately 470–790 MHz) and is often known as "white space" access. To the unlicensed user the band broadly appears to be just another unlicensed band, albeit there is some risk that there will be no spectrum available in particular locations or at particular times.

9.4.2 Advantages and disadvantages

Compared to other approaches, unlicensed shared access can be relatively simple to introduce as there is no need for any competition or auction to decide who can use the band. There is no need to clear out any existing user, only to determine the rules of sharing such that they will not suffer interference. Harmonization with other countries can be relatively easy to achieve since all can open up a piece of spectrum for white space access – even if the band is not exactly the same, the database will seamlessly advise devices as to the channels available in that country.

Unlicensed access also brings all the standard advantages of unlicensed spectrum, such as the ability to stimulate innovation, the speed of introduction of new services, and the ability to try a wide range of different applications.

9.4.3 Managing interference

Interference with licensed users can occur via two mechanisms:

1. Emissions from DSA devices that fall into the band used by licensed users. These are typically out-of-band emissions from DSA devices that fall in-band for the licensed device.
2. Emissions from DSA devices that are outside the band used by the licensed user, but which the licensed user's device is unable to filter adequately and hence result in interference. These are typically the in-band emissions from the DSA device that fall in channels close to that used by the licensed device.

In practice, a combination of these two will occur. For example, a DSA device operating two channels away from a TV receiver might cause interference as a result of its out-of-band emissions at $n+2$ channels (where n is the channel the DSA device is operating on). Simultaneously, it might cause interference as a result of its in-band emissions being poorly filtered by the TV receiver with limited rejection at two-channel separation. Which of these is most significant will depend on the relative performance of the transmit filters in the DSA device and the receiver filter in the TV (or other licensed device). For a given device location, if a database knows (1) the possible location of licensed receivers, (2) the frequencies they are using and their receive power levels at those frequencies, (3) the performance of the licensed receivers, and (4) the emission mask of the DSA device transmitter, then the database can determine the maximum transmit power that the DSA device can use before it causes interference.

The location of licensed receivers can be found through predicting the licensed-service signal strength and establishing coverage contours showing the areas within which the licensed service could be successfully received. Typically these are calculated using propagation modeling tools.

The TV frequencies in use can be sourced from the licensed user. In some cases, such as TV transmission, they are fairly static, but for other networks they may change even on a sub-second basis if techniques like frequency hopping are in use.

The performance of the licensed receivers can be obtained through specifications or, better, from measurements with real-world interference. Finally the emissions mask of the transmitter can be obtained from the device manufacturer, standards, or measurements.

The actual interference level experienced is given by:

$$I = P_{IB} * PL + \sum\nolimits_{\text{all other frequencies}} P_{OB}(f) * PL * FA(f)$$

Where P_{IB} is the power emitted in the band used by the receiver and P_{OB} is the power emitted outside this band at frequency f. In practice the dominant out-of-band effect is in the band used by the transmitting device and the equation simplifies to:

$$I = P_{IB} * PL + P_{OB} * PL * FA$$

The path loss comprises:

$$I = G_{TX} + G_{RX} - CL - SL$$

Where G_{TX} is the gain of the transmitting antenna in the direction of the receiver, G_{RX} is the gain of the receiving antenna in the direction of the transmitter, CL are cable losses or similar (e.g. as might result from a TV antenna on a roof using coaxial cable to feed into the house), and SL is the signal loss. The level of signal loss and how it can be modeled is described in section 2.3.

Once this information is complete, white space availability in a particular location can be found by drawing a coverage contour around the DSA device transmitter, operating at its specified transmit power level, out to the range where its transmitter power falls below the point where it could cause any interference to the licensed service and then for every "pixel" within this coverage area checking the spectrum availability. This is done by looking across a range of frequencies above and below the one being tested and seeing whether the combination of out-of-band emissions from the DSA device and the imperfect adjacent channel selectivity of the licensed receiver will result in interference. If so, that frequency is removed from the search process. Any remaining white space channels can be used up to the specified DSA device transmitter power level.

The number of channels returned can be improved through (1) reducing the DSA device transmit power and so decreasing the range over which it can cause interference and (2) reducing the DSA device out-of-band emissions, which reduces the signal levels falling in-band to the licensed service. Note that reduction in the out-of-band emission levels is only beneficial up to the point where the in-band emission from the DSA device becomes the dominant interference case. So, for example, with typical TV protection margin ratios, once the DSA device transmitter's out-of-band emission levels in a channel one away from the carrier ("$n + 1$") fall below 50 dBc (50 dB below the wanted signal), interference will be dominated by the in-band signal from the DSA interferer and further improvements in these out-of-band emissions will not deliver any further gains in white space availability.

9.4.4 Preventing interference between unlicensed users

Assuming that interference to the licensed service is avoided, there is still a risk that multiple DSA users will access the spectrum in the same place and could cause each other interference. This is a problem common to all unlicensed bands, such as that at

2.4 GHz, but with a new access method there may be new possibilities to control or avoid it. Options for unlicensed management of the band include:

1. leaving it to users to deal with the interference, as occurs in other unlicensed bands;
2. publishing codes of conduct that must be adhered to when accessing the band, such as the use of power control and maximum duty-cycle rules;
3. restricting the use of particular bands to certain technologies or applications such that coexistence is eased and likely band utilization lower; and
4. using the database to manage the band in some manner.

For the first case, users can decide what measures they wish to take. As a minimum they might select radio technologies that are robust to interference such as frequency hopping and strong error correction. In some cases, different standards bodies might work together to reduce the interference between their technologies. This happened, for example, when Bluetooth was seen to cause interference to Wi-Fi. The Bluetooth specifications were amended so that when Bluetooth devices detected a Wi-Fi transmission they adjusted their frequency-hopping pattern to avoid direct interference with it. This minimal interventionist approach from the regulator has generally worked well. It has enabled innovation by not restricting usage, and the cases of interference have generally been limited and successfully resolved without the regulator. Any deviation from this approach should show clear benefits before being adopted.

Ofcom have considered the use of codes of conduct [7], but noted the difficult balance between having rules that are effective and at the same time not restricting new technologies or approaches. In the end, they concluded that some "politeness protocols" could be used which provided high-level guidance as to how to avoid interference. Broadly these were:

- to use power control wherever possible so that the radiated power levels were minimized,
- to avoid continuous transmission such that others had a chance to gain access to any channel, and
- to group together like applications in terms of power levels and duty cycles since similar applications tend to coexist better than highly dissimilar ones.

The last of these touches on the third option set out earlier – restricting access to bands to particular applications. For example, a particular DSA band such as TV white space might be restricted to machine-to-machine (M2M) applications. This makes access for the M2M systems much more certain but blocks access for others. With no certainty as to which applications will be the most successful it is a risky and difficult decision for a regulator to make. Better, perhaps, to allow multiple applications and then encourage the most successful to find a licensed band that they can make their permanent home.

The final option is new to unlicensed spectrum access since databases have not been used in the past. For example, a database could note multiple requests for frequencies in the same area over a short time period and start to refuse some on the basis that congestion might occur. Or it could accept a certain number for one channel and then start giving out the next channel. However, all these approaches are very inexact since little is known about the level of usage behind each request and the actual congestion that is occurring.

One of the great advantages of a database approach is that policy can be changed over time without having to recall devices. Regulators could adopt a relatively open policy at first to see which applications emerge and then, if interference between DSA devices becomes a problem, they could progressively tighten regulation as appropriate at the time. This "wait-and-see" approach could be a very powerful new way of managing spectrum.

9.5 Advantages and disadvantages of shared access

On the positive side:

- access is free or low cost and does not require a license,
- prime frequency bands can be available.

However, there are some disadvantages:

- generally there is no certainty of access,
- there is a possibility of interference from licensed and other unlicensed users,
- bands may not be harmonized across multiple countries, and
- the requirement for geo-location and an alternative communications channel to contact the database (in all current implementations) can add to device cost and complexity.

This can lead to a dilemma in terms of the most likely applications to use shared access. On the one hand, DSA is (currently) unlicensed and the conventional wisdom is that wireless networks need licensed spectrum; hence DSA is not suitable for network deployments. On the other hand, the need for a geo-located master device with an alternative communications channel biases applications towards network solutions where the base station can readily fulfil these functions. How can this be resolved?

Broadly, the rules for geo-location, at least as they are currently implemented, tend to favor networks. This is because the "master device" which initiates any wireless communications needs to be able to locate itself and then use a non-white space channel to access the database. If this were used for a peer-to-peer application such as an enhanced Bluetooth-like link then the master, which might be a simple consumer device, would need GPS (which might not work if the device was indoors) and a cellular data link. These items would add substantial cost to the device. For a Wi-Fi type of application, white space might be practical. The Wi-Fi base station would need to locate itself, but it might be possible for the owner to self-declare the location or for the manufacturer to be able to build GPS into the device. Backhaul connection should be available. But simplest of all is a network where the base station is fixed and its location is known by the

> **Box 9-1.** Game theory, the commons, and dynamic spectrum access
>
> Much intellectual effort has gone into applications of game theory to dynamic spectrum access. Game theory studies the behavior of decision makers whose decisions affect one another. Firms or other organizations seeking access to spectrum clearly fall into this class. Games fall into two categories: co-operative and non-co-operative. A game is co-operative if agreements or commitments are fully binding and enforceable, and non-co-operative if they are not.
>
> This distinction is illustrated in Table 9-1 with respect to primary users (PUs) and secondary users (SUs).[6] Thus opportunistic use of white spaces is an entirely different mode of access for a secondary user than licensed spectrum access.
>
> **Table 9-1.** Categorization of spectrum access modes
>
> | | Non-co-operative | Co-operative |
> |-----------|------------------|--------------|
> | **Primary** | License-exempt | Spectrum license trading |
> | **Secondary** | White spaces | Licensed spectrum access |
>
> A key concept of game theory is the Nash equilibrium, a configuration of outcomes leaving no player with a desire to change his or her strategy, where that strategy may be either "pure" (certain), or "mixed" (probabilistic).
>
> A large number of particular models have been developed to elucidate the expected behavior of players in different conditions. For example, a group of authors has investigated how, rationally, firms would try to organize themselves co-operatively into coalitions of secondary users which share sensing information and thereby increase their transmission capacity.[7]
>
> Although this work has not yielded a major breakthrough in the organization of dynamic spectrum access, which has instead taken a more directive and regulated approach, it has encouraged rigorous thought about the structure of the various problems which have emerged.

installer. Backhaul would also be permanently available, making it simple to query a database.

Applications currently suggested for DSA tend to be network-based. They include fixed broadband to the home, mobile broadband, M2M, and perhaps military usage.

If most or all applications in a particular DSA band are network-based, this would enable co-ordination between the different network operators, helping to avoid self-interference between the unlicensed DSA users. This could help both improve the capacity of DSA bands and reduce the risk of increased interference from unlicensed users as the band gets more heavily used.

---

[6] Following [8]. [7] For this and other examples, see [9].

9.6 Example 1: TV white space

9.6.1 Why TV band?

The UHF TV band has become the preferred candidate for the first deployment of white space access.[8] This is for a range of reasons, as follows:

- **Perception of a lot of white space** The TV band extends approximately from 470 to 790 MHz, some 320 MHz of spectrum, with the perception that over 100 MHz is free in some areas.
- **Valuable spectrum** At these frequencies, signals propagate a long way, making it ideal for a range of applications.
- **Static licensed usage** Much of the usage is TV transmitters, which are stationary and rarely change their frequency or power levels; hence the database input is mostly static (although, as we will see in the next chapter, the program-making usage is more dynamic).
- **Licensed user does not own the white space** For many licenses, such as cellular, the license entitles the owner to sole usage across the country, so any white spaces belong to them. However, in most countries broadcast licenses are for single transmitters and hence any unused spectrum between transmitters is not owned by a licensee and so can be readily enabled for white space by the regulator.

While these factors are valid, there are a number of issues with the TV band that make sharing difficult. These include:

- **Receive-only nature of devices** TV sets do not transmit and so are impossible to detect, making sensing highly problematic.
- **Very high-power transmissions** TV transmitter sites operating at 100 kW and above can cause interference problems to DSA devices.
- **Changing band plan** TV bands are being progressively compacted, with the dividend being auctions at 800 MHz and progressively at 700 MHz.

9.6.2 Issues with broadcast sharing

In order to determine whether there is noticeable interference, the carrier-to-interference (C/I) ratio needs to be determined at the TV receiver, where the carrier is the wanted TV signal and the interference the signal coming from the DSA device. The carrier depends on factors such as the distance from the TV transmitter and the antenna in use. The interfering signal comprises two parts, the in-band element, which is the interference within the same frequency channel as the TV signal, and the out-of-band element, which is the signal in adjacent channels. The in-band element passes directly into the TV decoder circuitry while the out-of-band element is attenuated by filters in the TV receiver. The amount of attenuation will vary from manufacturer to manufacturer. The strengths of both elements of the

---

[8] This section is adapted from [10].

interfering signal depend on the transmit power of the DSA device and the propagation loss between the device and the TV receiver. Finally, the TV will be able to tolerate a certain level of interference depending on the signal level of the carrier, the waveform pattern of the interference and the encoding used on the TV signal.

So there are many factors to take into account; to summarize, the following are needed:

1. the signal level of the wanted TV signal, or carrier, at the TV antenna;
2. the gain of the antenna and loss of the cable before the signal arrives at the TV receiver;
3. the transmit power of the DSA device;
4. the output mask of the DSA device, showing the in-band and out-of-band power levels;
5. the waveform of the DSA device;
6. the propagation loss between DSA device and TV antenna;
7. the efficiency of the TV receiver filter in removing out-of-band interference; and
8. the C/I needed for the particular TV "mode" (mix of modulation level and coding) in use, receive power level, and interferer waveform type.

Some of these (factors 7 and 8) vary from TV to TV. Others (factors 1 and 2) vary from house to house. Some factors can only be predicted based on statistical models (factors 1 and 6). Deriving a clear and definitive assessment of interference levels turns out to be far from easy. We discuss each of these in the remainder of this section.

9.6.2.1 The TV signal level at the antenna

The TV signal level depends on the transmit power and the propagation loss between transmitter and household antenna. The transmit power is generally well known and fixed. The propagation loss depends on the distance between transmitter and receiver and the terrain in between. Prediction of this loss is made using propagation models such as the "Hata model" [11]. Propagation modeling is a complex area that can readily fill an entire book [12] – the key issue here is that models are only approximations. A model typically divides the country into "pixels" for modeling purposes. A pixel might be a square 100 m × 100 m. It then predicts the average signal level in this pixel. Because of inaccuracies in the model and fine detail in the terrain, such as large trees in front of the receiver, it is recognized that the signal level across the pixel will vary from this predicted average. Typically it is assumed that there is a log-Normal distribution of signal with a σ of around 5 dB.[9] There is also a time-varying element. TV signals can be affected by weather conditions. When there is high pressure over the transmitter an effect known as "tropospheric ducting" can take place which "tunnels" radio signals through the atmosphere, allowing them to arrive at destinations much

---

[9] A log-Normal distribution has the characteristic that 99% of the data falls within ±3σ of the average. So in this case almost all receivers can be expected to see a signal at least 15 dB below the average.

further from the transmitter than would normally be the case. This can be a particular problem for interference to a TV receiver from a far-distant TV transmitter using the same frequency.

For these reasons TV reception is often quoted for a particular time and location probability. So, a pixel might be declared to have 10 dB C/I at a 99% location and 50% time probability. This would mean that 99% of the receivers in the pixel would have a signal level of at least 10 dB C/I for at least 50% of the time.

This complicates calculating the impact of DSA interference since it cannot be simply said that it will, or will not, cause interference in a pixel. Instead, it will reduce the percentage of homes in that pixel that are able to receive TV signals, perhaps from 99% to 98%. In this case it would be said to have lowered the location probability by 1%. If there were only 10 homes in the pixel then this might have no impact, or might tip one of the homes into interference.

Hence, something as apparently simple as the TV signal level turns out to be a complex estimation represented with probabilistic models.

9.6.2.2 The gain of the antenna and cable

The TV signal just outside the house is not the same as that seen at the input to the TV itself. Most viewers use rooftop-mounted directional antennas that have gain. This improves the quality of the TV reception as long as the antenna is pointing in the right direction. There is then a cable between the antenna and the TV which will cause some loss in the signal. For modeling purposes it is often assumed that TV antennas have a gain of about 12 dB and that cable losses are some 3 dB, leading to a total gain of 9 dB.

Even this, though, turns out to be a simplification. Broadly, viewers install antennas that are just about good enough for the job. So where the signal level is high, a poor-quality antenna (or none at all) is often used. Even where a good-quality installation is in place, this typically degrades over time. In weak signal areas this would lead to loss of picture and so be rectified. In good-quality signal areas it is not noticed. A survey in the UK for a government department [13] showed gains varying from over 30 dB to −30 dB. There was a clear negative correlation between TV signal level and antenna gain. The net impact of this is that TVs will generally be operating towards the lower end of their input signal level – an important point that we will return to later.

9.6.2.3 The transmit power and output mask of the DSA device

This is more straightforward. The transmit characteristics are typically constant for a given device model and can be characterized by the manufacturer or some type-approval entity. However, these parameters will vary between different types of DSA device, such as an M2M terminal compared to a Wi-Fi router. Hence, an optimal implementation of DSA would know the characteristics of each device type and provide allocations of spectrum accordingly. This is what Ofcom in the UK are proposing but not a route taken in the US by the FCC, who treat all devices as having the same characteristics (which the FCC specify).

9.6.2.4 The waveform of the DSA device

In the early tests of the impact of DSA interference on a TV receiver, the interference was modeled as a constant, or "continuous wave" (CW) signal. This is the easiest signal to generate and was assumed to be sufficient to understand the impact.

It was only later in the process that real DSA device signals were used. A typical DSA device will not transmit constantly. Instead, it will send its message and then stop transmission until the next message. Even when transmitting it may well have a "burst-like" structure to its signal since most modern communications technologies make use of some form of time division of resources to divide the access to the system amongst multiple devices. The net result is that a more typical interference is an on-off-on-off type of signal that is often described as "bursty" rather than "continuous."

It was expected that this would be less problematic for a TV receiver since there is less energy in a bursty signal than in a continuous one (because there is no energy in the gaps). However, measurements showed that bursty signals actually had greater impact on TVs than continuous ones.[10] Worse, the variation was dramatic between different manufacturers of TVs, with some being almost unaffected and others having between 10 and 20 dB greater impact.

The reason for this appears to be the manner in which the automatic gain control (AGC) system within the TV is implemented. When it sees a strong interfering signal it reduces the amplifier gain of the receiver to avoid overloading it, often in a large step change. This step change destabilizes the rest of the receiver, which can take a fraction of a second to realign. When the interferer is constant, this behavior is perfectly fine. But when the interferer is bursty the AGC is forever stepping down and up, constantly destabilizing the rest of the receiver and destroying the picture. It would be possible for TV manufacturers to redesign their receivers so that they moved in smaller steps or took a longer-term average view of interference, and indeed many appear to be doing just this. However, that does not help the many millions of TV sets currently deployed.

In another twist to the story, it may be that the burst issue is less of a problem than anticipated. The tests on TV receivers tended to be performed in a laboratory with two signals injected into the receiver – a "pretend" TV multiplex on one channel and a DSA device on another. In this case the burst signal might make up 90% of the total signal energy received by the device and so have a major effect on the AGC. But in a typical real-world deployment a TV receiver might see four to six TV multiplexes and a range of interfering signals ranging from a nearby mobile tower to multiple DSA devices. Here the bursty signal might only be 10% of the total energy received and so have much less impact on the AGC. This is still an area of research, but this whole saga illustrates the complexities of interference and the need for careful real-world testing.

---

[10] See, for example, Figure 4.1 in the Bute Trial report [14].

9.6.2.5 The propagation loss between DSA device and TV antenna

Not all the signal emitted by the DSA device reaches the TV antenna. The signal reduces according to the distance between the device and the antenna, and according to whether there are obstacles and so on. It can also be attenuated by the directionality of the TV antenna if it is not pointing directly at the terminal, and by its polarization if the TV signal is differently polarized to the DSA signal.[11]

There are two different cases to consider:

- co-channel emissions, which typically only occur when the DSA device and TV receiver are some distance apart, and
- adjacent-channel and out-of-band effects, which can occur when the device and receiver are close together.

The location of both devices is generally not known in detail. As discussed earlier, modeling makes use of pixels to be tractable. The size of a pixel can vary from model to model and from time to time; at present most use 100 m × 100 m. When the DSA device reports its location, the database can determine whether it is in the same pixel as a receiver or not.

In our first case, of co-channel emissions, the DSA device has been allowed to transmit on the same channel as a TV receiver some distance away. Given the protection required for the TV receiver, these two devices would certainly not be in the same pixel, but likely many kilometers or tens of kilometers apart. In this case, relatively standard propagation models can be used, such as the Hata model, to predict the loss between the two devices. To be conservative, a number of standard deviations (e.g. 3σ) might be taken off the predicted propagation loss (making the interference signal stronger) before performing the test to see whether interference will occur. However, while such conservative choices do ensure a very low likelihood of interference, care needs to be taken not to become overly conservative.

In practice, it is not the co-channel emissions that dominate the availability of white space but the impact of adjacent and out-of-band effects. These are caused by two phenomena:

- out-of-band emissions from the DSA device, typically at some 50 dB less than the wanted emissions, falling into the TV band as much weakened co-channel interference, and
- the TV receiver not fully filtering the wanted DSA signal, hence some of the signal in adjacent TV channels appearing as interference.

In an ideally designed solution these two effects would be of about the same magnitude and the TV would see the combination of the two. These interfering signals are much

---

[11] TV signals are generally horizontally polarized (other than repeater stations). DSA signals are somewhat random but many base stations will use vertical polarization. Hence, some discrimination can be expected, at least between base station and TV receiver.

weaker than the co-channel case, often enabling DSA devices to be in the same pixel as a TV receiver. At this point, general propagation models cannot be used since the two devices might only be meters apart. Instead, assumptions have to be made as to the "minimum coupling loss" (MCL). This is the smallest reduction in the signal from the DSA device that might occur when the geometry and distance between the DSA device and TV antenna are such as to minimally attenuate the interfering signal. Because these adjacent-channel effects tend to be dominant, and because the database assumes a worst case based on the MCL, then determination of the MCL itself becomes critically important.

9.6.2.6 The C/I needed by the TV

The TV needs a particular C/I ratio both on the channel it is trying to receive and on neighboring channels for some frequency span above and below the wanted channel (often of the order of ±9 channels). On the wanted channel the C/I ratio is typically positive (so the TV signal needs to be greater than the interference) whereas on other channels it can be negative (so the TV signal on the wanted channel is weaker than the interfering signal on a neighboring channel). The reason the C/I ratio can be negative is because it is then attenuated by filters within the TV such that it becomes much smaller. The collection of C/I values is often termed the "TV protection ratios."

However, as discussed earlier, there are many factors that impact these protection ratios, including

- the strength of the TV signal – often TVs need greater protection ratios when the TV signal is strong;
- the type of interfering waveform, whether bursty or continuous, as discussed earlier; and
- the quality of filtering within the TV.

Variation between TV receivers is particularly problematic. The DSA database will not know which models of TV receiver are in use in any location and so may have to assume the worst-case receiver (which would be the one most susceptible to burst waveforms and/or with the least filtering). This can build a further tier of conservatism into the assumptions.

These factors can also vary over time as new models of receiver are introduced. It might be hoped that they would get progressively better at rejecting interference, but this is not always the case, especially if some new lower-cost solution becomes available that allows the manufacturer to reduce the price of their TV.[12] This suggests that there might be merit in specifications for minimum performance for receivers, which is something regulators have considered for

---

[12] For example, the gradual introduction of receivers based on silicon technology rather than discrete RF components has tended to worsen the performance of receivers, albeit allowing them to be much lower in cost.

180 Dynamic spectrum access

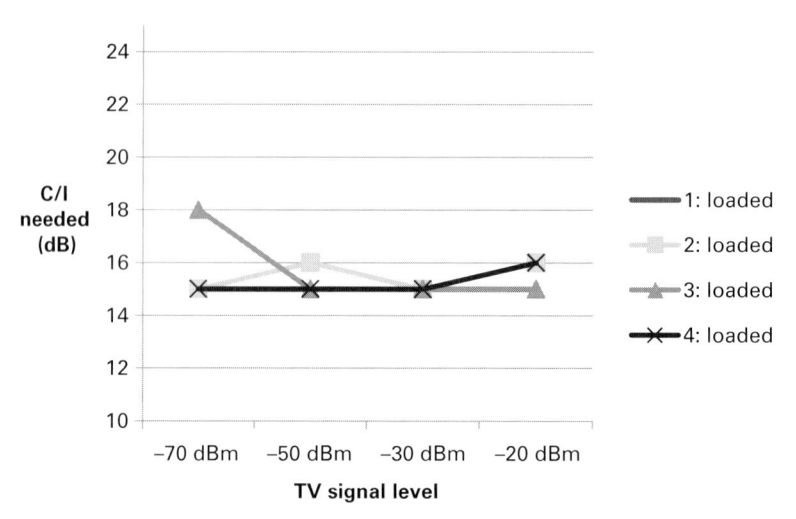

Figure 9-2. TV protection ratios from a fully loaded co-channel LTE transmitter. Source: [10].

many years but typically decided to be too difficult or disproportionate to implement.[13]

Many measurements have been made of the TV protection ratios needed – some examples are provided below.[14] These show the carrier-to-interference ratio needed for the TV to operate successfully for four different TV receivers and at four different received TV signal levels, ranging from the weakest signal that gives good reception (−70 dBm) to a very strong signal (−20 dBm). Note in these charts that in some cases some lines are identical and are not always visible on the charts, nor is there need to link a line to a particular receiver, the relevant issue being how closely they lie together.

Figure 9-2 shows four different TV receivers (labeled 1 through 4) in the case of the LTE transmitter being co-channel to the TV receiver and operating in a "loaded" mode such that its transmissions are continuous. In this case the TV signal needs to be some 15–20 dB above the LTE signal for successful reception. There is little variation between the four TVs and across the received signal level, which is broadly what would be expected.

Figure 9-3 shows the same situation but for cases of both a loaded LTE signal (solid lines) and an unloaded or idle base station (dashed lines) generating bursty transmission. In the idle case, around 5 dB more protection is needed to protect the TV reception.

Figure 9-4 turns to the case where the LTE transmitter is located nearly three TV channels (18 MHz) away. In this case the LTE signal can be stronger than the TV signal since the TV can filter much of the LTE signal. The figure shows somewhat similar performance across the TV receivers, albeit with 10 dB variation. But more importantly there is now a large variation with received TV signal level.

---

[13] In particular, the issue here is the ability of a TV to reject a non-TV waveform unknown at the time the TV was designed.
[14] These are based on measurements made by ERA on behalf of Ofcom [15].

TV white space 181

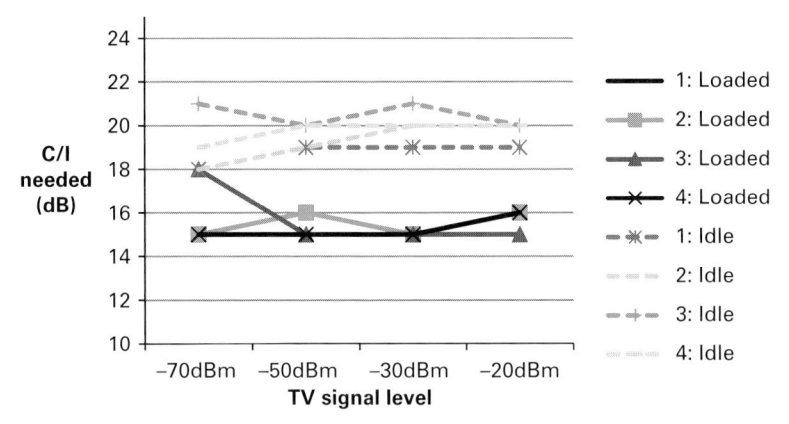

Figure 9-3. TV protection ratios from a fully loaded and unloaded (idle) co-channel LTE transmitter. Source: [10].

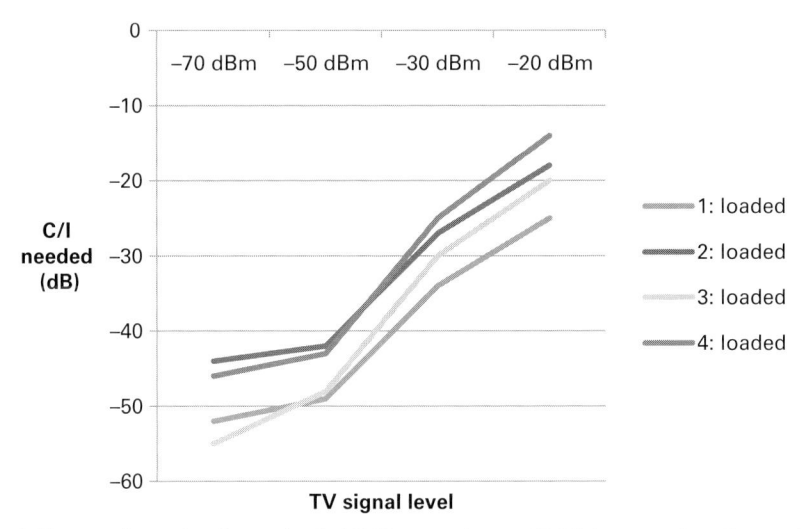

Figure 9-4. TV protection ratios from a loaded LTE transmitter at 18 MHz separation. Source: [10].

At low level, the TVs can tolerate the interferer being around 50 dB above the TV signal, but at high signal levels they can only tolerate around 20 dB. This is probably due to the LTE signal starting to overload the front end of the TV receiver and leading nonlinear responses in its receive chain. The net impact of this is that as the TV signal gets stronger, the interferer must stay at broadly the same level.

Figure 9-5 adds into the previous figure the case of bursty interference. This leads to some dramatic results. All receivers are at least 10 dB worse, but receiver 4 is spectacularly bad, some 25 dB worse than the continuous case.

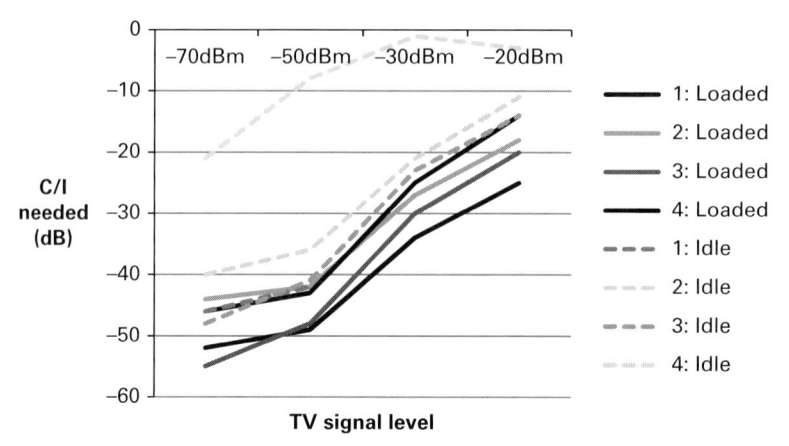

Figure 9-5. TV protection ratios from both a loaded and an unloaded (idle mode) LTE transmitter at 18 MHz separation. Source: [10].

If a regulator were to protect the worst receiver in the worst case then at 18 MHz separation the C/I ratio would be around −1 dB. If they were to protect the best receiver in the best case then it would be around −55 dB. This is a massive difference requiring difficult judgment from the regulator.

However, as discussed earlier, if the assumption is that TV receivers always operate toward the low end of the received signal level due to "just good enough" antenna and that bursty signals are less problematic in the real world due to the presence of many other signals, then the candidates become just the loaded transmission at −70 dBm or thereabouts. There is much less spread here, making the regulatory decision simpler.

9.6.2.7 Minimum coupling loss

The MCL is the smallest amount of attenuation that can be envisaged between the DSA device and the TV receiver. In order to determine this loss it is first necessary to predict the worst-case geometry when the DSA device and TV receiver are at their closest. This requires some judgment. For example, the closest possible would be when the DSA device was being held within touching distance of the TV antenna. But with TV antennas mounted at rooftop level this is highly unlikely. As well as physical closeness, the directionality of the TV antenna has a large part to play. For example, a DSA device could be in an upstairs room with the TV antenna in the loft space. They might only be a couple of meters apart. But there would be a large attenuation of the signal both as it passed through the ceiling and because it would be attenuated by the antenna since it would be arriving at a direction where the TV antenna had significant negative gain.

The worst, somewhat practical, situation would be if there was a house across the street that was taller than the house with the victim receiver. If the antenna pointed directly at an upper floor on this house and there was a DSA device on a balcony directly opposite the victim then the separation distance might only be the width of the road (say six meters) and the full gain of the antenna might apply. In this case the path loss could be as low as 32 dB with the result that the emissions from the DSA device are only somewhat attenuated and could cause interference. Allowing for random orientation of the TV antennas, then the MCL has a distribution for which a log-Normal (49,8)[15] curve can be shown to be a reasonable fit. But how likely is this situation in practice? If nothing else, the TV reception would be unlikely to work even without the DSA device since the signal would be blocked by the house across the street.

9.6.2.8 Indoor TV

A key question in the protection of TV receivers is whether indoor reception – that is, using an antenna on top of the TV – is to be protected. Indoor reception changes a number of the factors set out above, including:

- the antenna gain is much lower and less directional,
- the MCL calculation is quite different since the DSA device could be much closer to the TV antenna, and
- the TV is often operating much closer to the margin of minimum TV signal and so might be more readily impacted by interference.

In the worst case, protecting indoor TV can essentially remove all white space access. This is because, with a very low MCL, even a very low-power signal from the DSA device can cause interference. In practice, regulators take a variety of approaches that tend to soften this.

Some regulators do not protect indoor reception.[16] In their view, TV licenses were provided on the basis of providing coverage to rooftop antennas and if viewers happen to be able to use an indoor antenna that is fortuitous but they cannot expect that to be protected.

Others take the view that the MCL should not include devices in the same room, or even the same building, as the indoor TV. That is because both the DSA device and the TV receiver would be under the control of the same person and they might be able to do their own form of "interference management" by turning one off, or moving them apart. (In practice, they may be unaware of which DSA device is causing the interference.) In this situation the MCL is then calculated for a device in an adjacent building with typical separation of 10 meters or so and also the penetration loss associated with the signal passing through a wall. Most calculations show that the MCL for this case is very similar to that for the case of the outdoor antenna described earlier.

9.6.3 Current status

At the time of writing the number of white space deployments was limited. No significant interference cases were known, but equally the density of deployment was insufficient to draw any conclusions from this. However, equivalent experience with

---

[15] That is, a log-Normally distributed set of data with a mean of 49 dB and a σ of 8 dB.
[16] See para 5:13 in [16].

LTE suggested that the combination of risk-averse factors was likely to result in significant overprotection.

9.7 Example 2: US 3.5 GHz band

9.7.1 US proposals

The FCC in the US in 2013 put forward some novel proposals for the use of the 3.5 GHz band – a band predominantly used at present by government (defense) users [17]. They started from the viewpoint that small cells with a range of 100 meters or so are important to cover targeted indoor, or localized outdoor, areas ranging in size from homes and offices to stadiums, shopping malls, hospitals, and metropolitan outdoor spaces. They noted that small cells can be deployed relatively easily and inexpensively by consumers, enterprise users, and service providers, and hence a different regime to fully licensed might be appropriate.

The FCC felt that the 3.5 GHz band was well suited to small-cell deployments. The higher frequency was not an impediment because the range requirements were low and there was substantial spectrum available at these frequencies. Incumbent uses in the band include high-powered Department of Defense (DoD) radars as well as non-federal Fixed Satellite Service (FSS) earth stations for receive-only, space-to-earth operations, and feeder links. This leads to large "exclusion zones," which cover approximately 60% of the US population, hence the 3.5 GHz band would not be particularly well suited for wide-area macrocell deployment.

The FCC proposed three tiers of service, as follows:

1. Incumbent Access,
2. Priority Access, and
3. General Authorized Access (GAA).

The Priority Access tier would consist of a portion of the 3.5 GHz band designated for small-cell use by certain critical, quality-of-service-dependent users at specific, targeted locations. These could include hospitals, utilities, state and local governments, and/or other users with a distinct need for reliable, prioritized access to broadband spectrum at specific, localized facilities.

In order to prevent an expectation of quality of service in areas where such an expectation might not be warranted, Priority Access operations would only be permitted in geographic zones with no likelihood of harmful interference from Incumbent Access users and no expectation of harmful interference from GAA users to Incumbent Access users. Priority Access users would be required to register and be accorded protection from interference from lower-tier users and other Priority Access users within their local facilities. This is shown diagrammatically in Figure 9-6.

The FCC proposes a spectrum access system (SAS) to govern operation within and among tiers. The SAS would incorporate a dynamic database enabled with geo-location to manage access across a number of planes, including geography, time, and frequency,

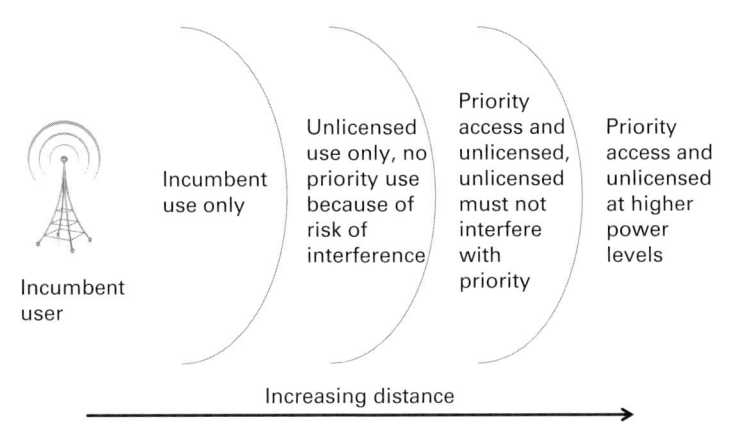

Figure 9-6. Sharing arrangements at 3.5 GHz.

and by other technological co-ordination techniques, modeled after the existing TV white space database requirements. The spectrum access system as applied to the 3.5 GHz band would cover some novel issues and might require a new generation of this dynamic database technology. The FCC felt that its creation would require significant planning and testing.

The key features of the proposal are:

- the use of a database to keep track of users, exclude certain geographies and protect higher-tier from lower-tier users;
- some parallels to white space with shared access into licensed spectrum;
- multiple tiers which try to mix a range of different usage types and availability requirements; and
- a prescriptive selection of applications – "small-cell."

The novelty is really in a mix of "LSA" and "white space" – i.e. both licensed shared access and unlicensed shared access at the same time and in the same band.

9.8 Example 3: government sharing

One area that has already started to be explored is the idea of sharing with governmental, or federal, users. Indeed, most regulators would note that sharing with government users has been in place for many decades – for example the 2.4 GHz unlicensed band is often shared with military use, as is the 5 GHz band in some European countries, where it is used for military radar. The early cellular systems shared some of their spectrum with the military, with cellular use in dense urban areas and the military elsewhere, before the military eventually relocated to other bands.

An influential report from the President's Council of Advisors on Science and Technology (PCAST) in the US was published in mid-2012 [18] which broadly suggested the extension of white space access techniques into governmental spectrum.

It recommended that 1,000 MHz of shared access be immediately identified to be made available on a three-tier sharing basis where the tertiary tier is a lightly licensed use of the spectrum. It suggested that this was the only practical way of gaining access to spectrum currently used for federal applications but also noted that this would provide a fertile ground for innovative wireless technologies and services. While generally welcomed, not all agreed with the conclusions and some expressed concern that pressure needed to be maintained on clearing spectrum as well as pursuing the alternative approached advocated in the report.

While the ideas in the report are not all new, and many have been voiced before, the timing of the report appears promising, fitting in well with the current issues and possibilities in the spectrum space. But the ideas in the report will only be implemented if they are "institutionalized." For example, the NTIA has had a testbed for sharing available for three years, but it has made little progress. As a result, calling for another testbed is unlikely to reap benefits unless a different approach is adopted.[17] In the National Broadband Plan [19] there was discussion of sharing but this had not been taken forward. This was not due to any lack of political will, but because there was no single person with the authority to implement ideas of this sort. Instead, decision making was dispersed with many veto rights enabling those not in favor to block progress.

Government and industry working together might overcome some of these problems and result in a more widespread understanding of the issues and data associated with federal systems.

Successful introduction of these concepts will require the incentives of those within government, and particularly within the NTIA, to be correctly aligned. At present, there is little incentive for those using spectrum to share it with others. Such sharing would add complexity, cost, and delay to their work for little apparent benefit to them. The classic incentive, used widely in the commercial world, is that of profit. Commercial users are assumed to have an incentive to share or sell underused spectrum because of the revenue that might flow to them as a result. It is well understood that this incentive is weak in regard to the government. Typically, if a department does raise revenue it is appropriated by the Treasury, either directly or as a lower budget settlement in subsequent years. Attempts to apply "incentive pricing" in the UK initially had mixed success. Direct monetary payments are unlikely to be useful. The PCAST report makes some suggestions as to other kinds of "artificial currency" that might be an alternative. But there remains a poor understanding outside government as to the pressures and incentives currently facing governmental employees and no clear solution to how to modify these such that appropriate decisions on spectrum were made.

One way to liberate federal spectrum might be for the government to make use of commercial systems where appropriate, thus enabling it to shut down its own systems. However, the evidence is that this was already done where possible and that for other systems, such as military radar, there was rarely an acceptable commercial solution

---

[17] The NTIA are pursuing the idea of a test facility which could be relatively quickly used when interference concerns surface. The facility might include a range of chambers and test ranges and flexible equipment allowing rapid real-world testing.

available. Hence it is unlikely that there were any "quick wins" here. Nevertheless, there is incomplete knowledge of the federal use of spectrum outside government. Closer interaction between government and commercial users of spectrum is a way to overcome this. Such interaction would not just facilitate shared understanding but would enable better provision of key data that would be needed when undertaking shared spectrum activities. This would apply also to discussions between the FCC and the NTIA, which were often not open to other commercial players. Making these discussions more open is important as part of the process of fostering a shared understanding of the issues. The database proposed by PCAST would hold much of this information and might be a good starting point for interaction and shared understanding. However, information such as the economic value of spectrum in alternative uses is missing from the proposals for this database and yet is a key enabler in making informed decisions.

More flexible licensing could help enable innovative new ideas, especially those that required different spectrum access methods. There has already been much work on enhancing flexibility over many years and the PCAST team gave serious consideration to this area. For example, in some cases sensing might be used alongside database access, particularly where updating the database was problematic for operational or security reasons. Sensing might also work alone, or with pre-stored versions of the database, in remote cases where peer-to-peer communication was needed but there was no ability to connect back to the database. This is a fertile area for research activities where new ideas and solutions might emerge and regulations should be such as to not prevent this.

Ofcom have also looked at governmental sharing, and while no explicit plans had been published at the time of writing, there was enthusiasm to explore this area further, with Ofcom staff commenting at conferences that the TV white space access might be a "trial" for eventual deployment of database access into spectrum such as that used by the government. Other regulators around the world can be expected to follow suit.

9.9 In conclusion: the need to increase flexibility

Spectrum management has evolved over a century or more. As the use of spectrum has grown it has become increasingly difficult to update and change spectrum management to reflect this. A simple parallel is "defragmentation." On a computer it is periodically useful to reconsolidate files into contiguous chunks and otherwise tidy up the disk. Spectrum usage is extremely fragmented and "defragmentation" would be very valuable, but is almost impossible because of the difficulty of relocating millions of users, many of whom would need new equipment. Similarly, licenses have been given out under various terms for ten or twenty years and cannot be changed until they expire.

This complicates the process of reform, but does not eliminate either its desirability or, in the long run, its necessity. It is therefore worth thinking about what a desirable destination for spectrum management might look like, and how we might seek to get there.

The defining feature of recent work on spectrum has been the focus on sharing. This is a natural consequence of the combination of the sharply increasing demand for spectrum in key bands, and the recognition that much spectrum, including the most valuable, is heavily underused. Spectrum has always been shared via limited commons in the geographic dimension, and in large temporal chunks (for example by day of the week). But new real-time technologies for sharing are now available which enable different users to respond dynamically to changing conditions of congestion in ways which were quite impracticable even a decade ago. Regulation, mostly at the national level, is catching up with this, via arrangements for overlays and underlays, "white spaces," licensed spectrum access, and so on. Public-sector spectrum, where the user is the government or a public body, is in principle most amenable to change in this regard (see Chapter 12), although public-sector users' appetite for spectrum, and their conservatism, should not be underestimated.

In these circumstances, the route to increased flexibility is likely to lie in the diminution of users' expectation of having exclusive use of spectrum. Only in exceptional cases should an organization expect to wholly own and control the spectrum which it has been awarded. The norm should be one of a variety of arrangements involving either a hierarchy of users' rights, with priority awarded to those of higher rank, or a single rank of users, with access based on a "first-come, first-served" principle (subject to regulatory limitations on any organization's scale of use), or a combination of the above.

This offers a variety of options which may be applicable to different bands and uses. In some, an operator might have primary rights of access, but other users should be allowed to work in and around these. Licenses should be structured to reflect this, with agreed levels of interference and mechanisms to resolve problems.

This degree of coexistence can be achieved through the currently available technology of geo-location database access, which should become the default approach for all devices, although peer-to-peer devices (which are mostly only connected to neighbors rather than to a network) may only be required to consult databases infrequently and when they are able.

In such a framework, there might appear less certainty of access for today's license holders. However, access rules can be constructed to provide as much prioritization as is felt necessary. The prices of access to spectrum would depend on the associated rights of access. The configuration of bands into different categories of user rights could be determined by the spectrum regulator or by a competitive auction process which would accept and compare bids for different levels of access, assigning access to the band to the combination of users' willingness to pay which yields the greatest revenues. As noted in section 5.6 above, a means can be sought to incorporate in this process the willingness to pay for *unlicensed* access to the band in question. A cap can also be placed upon the proportion of spectrum of any type which is in exclusive use. This would have an effect equivalent to the method of set-aside for new entrants which has been used in some spectrum awards – with the difference that in this case spectrum would be set aside for spectrum to be assigned for shared purposes. Once in place, negotiations between users of the band, where practicable, could alter the rights of access, with compensatory payments or similar made as needed.

In conclusion: the need to increase flexibility 189

We envisage that for a period the outcome would be a patchwork quilt of modes of spectrum access, but with a strong bias in favor of sharing, leading to an increase in spectrum utilization and decrease in spectrum prices. One possible outcome for certain bands would be a spot market for spectrum access of the kind envisaged several years ago by Eli Noam [20]. He forecast that users would gain access to spectrum on a pay-as-you-go basis, instead of gaining an exclusive award. Spot and futures markets would develop. This has become feasible because, since Noam's suggestion was first made, technologists and economists have made great strides in solving the interference problem by means other than exclusivity.

It is possible to see an analogy with the development worldwide of wholesale markets for electricity which successfully link retailers of electricity and their customers to a variety of heterogeneous electricity generators. In this case equalization of supply and demand is achieved on an instantaneous basis, via the operation of special balancing arrangements. In spectrum a variety of markets could coexist, differing in the degree of flexibility of access times required. The development in some jurisdictions of wholesale-only mobile wireless networks could be the precursor for such services [21].

However, there is no need to be overprescriptive about how such arrangements should develop, provided that the incentives for more efficient use of spectrum are put in place, and provided that spectrum users with market power are not able to foreclose entry into their markets.

What steps can be taken today to speed this process of transition? It would require action to be taken by different agents in the multilevel spectrum regulatory process, and we do not seek to spell it out in detail.

A useful first step would be the recategorization of spectrum. Instead of using terms such as "fixed," "mobile" and "broadcasting," spectrum needs to be divided according to power levels and duty cycles. Simplistically there might be high-power bands, medium-power bands and low-power bands.[18] It would be very helpful if the ITU and other supranational regulators made this switch.

Moving away from exclusivity could be a progressive process, accelerating as trust grows in the technical capacity of the databases or other mechanisms to protect users' rights, to avoid interference and to implement contractual or regulatory priorities. Some applications might be kept exempt from sharing for a period. But the option of sharing could be progressively introduced into other bands when they were re-auctioned or otherwise returned and repurposed. With some licenses having a long duration and others being effectively perpetual, the resulting extremely long timescales would be insufficient to encourage industry to invest in developing appropriate technology.

An alternative approach would be to set a timetable for the elimination of full exclusivity of licenses, perhaps with a different date for each key service (e.g. broadcasting, cellular) phased over a few years. (This was the approach adopted for the introduction of spectrum trading by Ofcom and others.) Multinational agreement on timescales and phasing would be helpful in stimulating the equipment market. Once the

---

[18] For further discussion, see Ofcom, Licence Exempt Framework Review.

process of eliminating exclusivity has begun, the greater availability of spectrum access would reduce the commercial benefits accruing to surviving exclusive licenses.

In such a framework, there might appear less certainty of access for today's license holders. However, access rules can be constructed to provide as much prioritization as is felt necessary and equipment design might be expected to evolve to handle more dynamic spectrum access.

The elimination of exclusivity is probably the most important, but not the only, measure available to enhance the flexible use of spectrum. Regulation and usage rights in spectrum, and the whole process of determining access rights, should be as much concerned with receiver performance as with transmitter performance – or, more precisely, it is the mix of transmitter and receiver parameters that the regulator needs to be concerned with.

We envisage that there would emerge from this process much more unlicensed and lightly licensed use spread across the spectrum. A wide range of different approaches should be available to encourage innovation. Some bands may need to transition from unlicensed to lightly licensed as the applications within the bands mature and need greater certainty of access.

A number of these areas are covered in more detail below, including sharing with governmental users, the specification of usage rights, and receiver standards.

References

[1] A. Tsertou and D. I. Laurenson, "Revisiting the Hidden Terminal Problem in a CSMA/CA Wireless Network" (July 2008) 7(7) *IEEE Transactions on Mobile Computing* 817.
[2] See http://stakeholders.ofcom.org.uk/binaries/spectrum/spectrum-policy-area/projects/ddr/eracog.pdf.
[3] See http://hraunfoss.fcc.gov/edocs_public/attachmatch/FCC-10-174A1.pdf.
[4] Ofcom, "Statement on Licence Exempting Cognitive Devices Using Interleaved Spectrum" (July 1, 2009), at http://stakeholders.ofcom.org.uk/consultations/cognitive/statement.
[5] See www.ict-cogeu.eu.
[6] RSPG opinion on LSA, RSPG13–529 rev1.
[7] See http://stakeholders.ofcom.org.uk/consultations/lefr/statement.
[8] M. Weiss, "Dynamic Spectrum Access" (2013) 37 *Telecommunications Policy* 193.
[9] B. Benmammar, A. Amraoui, and F. Krief, "A Survey on Dynamic Spectrum Access Techniques in Cognitive Radio Networks" (2013) 5 *International Journal of Communications Networks and Information Security* 71.
[10] W. Webb, *Dynamic White Space Spectrum Access* (2013), at www.webbsearch.co.uk/publications.
[11] See http://en.wikipedia.org/wiki/COST_Hata_model.
[12] C. Haslett, *Essentials of Radio Wave Propagation*, Cambridge University Press, 2008.

[13] See http://stakeholders.ofcom.org.uk/binaries/research/tv-research/aerials_research.pdf.
[14] See www.wirelesswhitespace.org/media/28341/tsb100912_bute_ws_report_v01_00.pdf.
[15] See http://stakeholders.ofcom.org.uk/binaries/consultations/949731/annexes/DTTCo-existence.pdf.
[16] Ofcom, "TV White Spaces: Approach to Coexistence" (September 2013), at http://stakeholders.ofcom.org.uk/binaries/consultations/white-space-coexistence/summary/white-spaces.pdf.
[17] See FCC 14–49, "Further Notice of Proposed Rulemaking (3650 MHz Band)," at www.fcc.gov/document/proposes-creation-new-citizens-broadband-radio-service-35-ghz.
[18] "Realizing the Full Potential of Government-Held Spectrum to Spur Economic Growth," at www.whitehouse.gov/sites/default/files/microsites/ostp/pcast_spectrum_report_final_july_20_2012.pdf.
[19] FCC, "Connecting America: The National Broadband Plan," 76, at www.fcc.gov/national-broadband-plan.
[20] E. Noam, "Today's Orthodoxy, Tomorrow's Anachronism: Taking the Next Step to Open Spectrum Access" (1998) 41S2 *Journal of Law and Economics* 765.
[21] See E. Flores-Roux, "Mexico's Shared Spectrum Model," at http://broadbandasia.info/wp-content/uploads/2014/04/EFloresRoux-Mexicos-shared-spectrum-model-March-2014-2.pdf.

10 Controlling interference
Licensing and receivers

10.1 Introduction

In previous chapters we have seen how one of the key functions of spectrum management is the control of interference. In Chapter 2 we looked at how interference occurred and towards the end of the chapter noted the difficulties in optimally regulating interference levels.

In this chapter we build upon an understanding of interference to look at how licenses might be structured so that they better define and control interference levels – enabling market mechanisms such as change of use without compromising nearby users of spectrum. This naturally leads on to looking at whether the performance of receivers in an interference environment requires greater specification.

10.2 Spectrum usage rights

10.2.1 Introduction

In a world where the use of a band of spectrum is unchanging, the regulator can determine various band parameters such as transmission levels and guard bands based on these specific uses and technologies. This is the approach that has been adopted for over a century. However, with increasing change of use of spectrum, e.g. from broadcasting to mobile broadband, this assumption is no longer valid. When the use of the band changes, the interference levels and patterns can change massively, potentially causing issues to existing users. If these existing users do not have some level of certainty about the interference they may suffer in the future (in some cases 10–20 years hence) then they may not invest in expensive infrastructure. Therefore, some solution to this problem that allows for change of use of spectrum but also provides certainty of interference levels is needed.

The actual interference suffered by a license holder will depend on the power emitted by the neighboring base station and the distance from that base station. A network of a relatively small number of high-power sites, such as that used by broadcast systems, would result in relatively few areas where interference might be experienced, but these areas would be relatively large. Conversely, a network of many medium- or low-powered sites, such as used by a cellular system, would result in many more areas where interference might occur, but these areas would be smaller.

There are a number of options to solving this problem:

- Simply set the maximum transmitter power under the assumption that the density of base stations is unlikely to change significantly. However, this may be an invalid assumption and adds risk to the deployment of networks by neighbors.
- Require every transmitter deployment to be agreed with neighbors. This would ensure that the neighbors controlled any change in interference, but would likely be restrictive, bureaucratic, and difficult to work in practice.
- Set the interference limits on the basis of the number and size of the areas where certain levels of interference could be caused. While this would not specify where those areas actually were, it would allow probabilistic design assumptions to be made.

None of these are perfect, but our preference is for the last. In this case an operator would be told that the interference it generates must not exceed a certain limit, x dBm/unit bandwidth, for more than a certain percentage of time, $y\%$, in more than a certain percentage of locations, $z\%$. In order to verify whether this was the case, a sufficiently large number of measurements would then be made across a defined unit area such as 5 km^2, for a long enough time to capture variations. In practice, modeling might be used rather than actual measurement to save time and cost.

Such an approach can be used both for out-of-band and for in-band emissions, since the interference levels experienced by a neighboring receiver are determined in an identical manner for both.

However, the use of such a set of rights leads to a different set of restrictions on the license holder than the existing approach of limiting maximum transmitter power. Under the current system a license holder typically has restrictions on what they can use the spectrum for, such as fixed or mobile communications, or another particular technology, but within those restrictions and up to a certain maximum power limit they can deploy as many base stations as they wish. If rights were restated in these new terms then there would be no restrictions on usage or technology. However, deploying a more dense network of base stations might only be allowed if the average transmit power from each base station was reduced (which is often the case in practice anyway).

10.2.2 Determining interference

Because a license holder knows the interference level it can expect, determining excessive interference is a case of measuring the distribution of received interference and determining whether it is in excess of that which neighbors are allowed to cause. This can be done using standard measuring equipment. There may be practical difficulties associated with time-varying or location-varying interference requiring longer-term monitoring.

Once a license holder has determined that there is excessive interference, it can report this to the regulator for resolution; alternatively it may, if it wishes, deal directly with the source of the interference. This is very similar to the position today. According to experience in Australia, the approach of dealing directly with the interferer is the one adopted most frequently.

There are two types of interference – illegal and legal. In the case of illegal interference, which cannot be settled by direct negotiation, the regulator should instigate the appropriate enforcement process. This can be accompanied by private legal action taken in the civil courts by the aggrieved party.

Legal interference is more complicated. It implies that the regulator has wrongly set the property rights for a neighboring license. In this case, the regulator needs to amend the license terms appropriately.

10.2.3 The UK SURs

The UK has studied property rights (which they term spectrum usage rights – SURs) in some detail [1], and has derived a proposed set of license terms. These are reproduced below.

For controlling emissions into neighboring geographical areas the following could be used: "the aggregate ... PFD at or beyond [definition of boundary] should not exceed X_1 dBW/m^2/[reference bandwidth] at any height up to H m above local terrain for more than P% of the time." For controlling emissions outside the license holder's frequency band (that appear as in-band interference for a neighbor) the following could be used: "the OOB PFD at any point up to a height H m above ground level should not exceed X_2 dBW/m^2/MHz for more than Y% of the time at more than Z% of locations in any area A km^2." For controlling emissions inside the license holder's frequency band (that may cause interference to neighboring users in frequency due to imperfect receiver filters), the same measure could be used: "the in-band PFD at any point up to a height H m above ground level should not exceed X_3 dBW/m^2/MHz for more than Y% of the time at more than Z% of locations in any area A km^2." Some possible numbers for these limits are also suggested by Ofcom, shown in Table 10-1.

10.2.3.1 An example case

In order to illustrate some of the points raised above consider the case where, due to the introduction of digital TV, the broadcasters no longer need their entire allocation at UHF. They decide to trade it to cellular operators who are looking for some spectrum at lower frequencies in order to improve their coverage of rural areas. It is likely that the property rights will not be ideal for cellular operators as they will be set to allow a few high-power transmitter sites rather than the multiple lower-power sites that the cellular operator might prefer. Prior to buying such a license, the cellular operator will wish to consult with their new neighbors, who may also be the sellers of the spectrum. We assume that the neighbor is willing to discuss changes to the property rights (otherwise the cellular operator would likely not proceed with their acquisition).

The downlink and uplink can be considered separately – if time division duplex (TDD) systems are used then the downlink and uplink are effectively within the same frequency band.

Table 10-1. Possible values for SURs

Aggregate in-band PFD at or beyond geographical boundary should not exceed X_1 dBW/m²/[reference bandwidth] at any height up to H m above local terrain for more than P% of the time	X_1 = (based on sensitivity of services in neighboring areas and any international agreements) H = 30 m AGL P = 10%
Out-of-band PFD at any point up to a height H m above ground level should not exceed X_2 dBW/m²/MHz for more than Y% of the time at more than Z% of locations in any area A km²	H = 30 m AGL X_2 = (based on service and standard "mask" for most likely technology also may be multiple values for different separations from band edge) Y = 10% Z = 50% A = 3 km²
In-band PFD at any point up to a height H m above ground level should not exceed X_3 dBW/m²/MHz for more than Y% of the time at more than Z% of locations in any area A km²	H = 30 m AGL X_3 = (based on service and maximum transmit power of most likely technology) Y = 10% Z = 50% A = 3 km²

Downlink. Cellular systems in large countries comprise thousands of base stations, whereas TV systems typically only have tens or hundreds. As a result, the transmit powers of the cellular system will typically be much lower than that for the TV system but users will likely be closer to a cellular base station than to a TV station. There are two potential areas of interference – from cellular to broadcast and from broadcast to cellular:

- Interference from the cellular to the TV system could occur when a cellular base station is close to a TV set. The worst case would be when a TV set is on the edge of the coverage of a TV transmitter and a cellular tower is erected close to their house and in a direct line to their TV antenna. In this case, both the in-band and out-of-band emissions from the cellular transmitter could be problematic. However, the property rights that the cellular operator inherits will likely require relatively low transmitter powers if there are many more cell sites than in the broadcasting case. These low power levels might be sufficiently low to prevent any significant interference, and indeed, more likely to prevent the cellular operator economically deploying a network. The cellular operator and the broadcaster might wish to study the maximum powers that the cellular operator can use without causing interference and modify their property rights accordingly. Or they might agree on deployment scenarios that minimize this likelihood.
- Interference from the TV to the cellular system could occur when a cellular user is close to a TV base station and so receives a high signal level. Because there are few broadcast base stations, and they are often somewhat remote, this will be a relatively infrequent occurrence. The cellular operator will have to accept this interference as part of the condition of buying the license. This will somewhat reduce the value of the

license to the cellular operator, who will not be able to provide coverage in the close vicinity of TV transmitters. The cellular operator might be able to negotiate with the broadcaster to reduce their transmitter power, but in practice this is unlikely as it would result in a loss of coverage to a broadcaster with universal service obligation.

Uplink. The interference is from the users of the cellular system to users of the TV system and from the TV transmitter to the receiver in the cellular base station.

Interference from the cellular system to the TV system could occur when a user of a cellular handset is near the antenna of a TV set. With most TV antennas mounted at rooftop level, this will be an infrequent occurrence. The most problematic cases will be when a user of a cellular phone is in the same room as a TV using a set-top-mounted antenna. The initial property rights will likely make mobile transmission quite difficult as the probability of exceeding the interference thresholds in the license, given a dense population of mobiles, would be high. Therefore, very low mobile transmit powers would need to be used, which would likely make the system uneconomic.

As a result, the cellular operator will wish to negotiate increased power limits. The in-band and out-of-band limits for the cellular phone need to be set to minimize the likelihood of this interference. However, some judgment needs to be used. If the limits were set such that there would be no noticeable interference, even were the handset placed next to the TV antenna, the emission limits might be so restrictive as to prevent the cellular system working properly. Instead, limits might be defined so that there was little noticeable interference when the handset was more than a meter from the TV antenna. Alternatively, guard bands could be used to separate the two applications further in frequency terms.

Interference from the TV system to the cellular receiver might be the most serious case of all. Here, signals from a TV transmitter are received by a cellular base station, which is trying to receive relatively weak signals from mobiles. Because both base stations might be mounted on hilltops, the propagation from one to the other might be good, resulting in a strong interfering signal at the cellular base station. Again for reasons of practicality and universal service obligation, the cellular operator will likely need to accept the increased power transmitted by the broadcaster. To mitigate the problem, the designer of the cellular system might select their cell sites to be as far away as possible from known TV transmitter sites, or insert filtering that can ease the problem.

The discussion above illustrates that there may be up to four situations of interference to consider – from system *A* to system *B* and from system *B* to system *A* in both the uplink and the downlink.

Having studied these cases, the cellular operator might enter into discussion with the neighboring broadcasters, putting forward some proposals as to how the property rights could be changed. The cellular operator might in compensation offer some payment or other concession to the broadcasters. Once these negotiations had concluded, the cellular operator could determine the type of system they could deploy and build a business case. This would allow them to understand what they could pay for the spectrum that they wish to purchase, and if viable move ahead with the acquisition.

10.2.4 Introducing SURs

To date, Ofcom have only issued SURs for one frequency band (the "L-Band" from 1,452 to 1,492 MHz). Outside this, there has been resistance to their introduction from license holders who see SURs as complex and untested. While this is true, the risk of interference from change of use is ever-growing. The US is now studying similar approaches (see below), and we expect to see SURs, or similar licensing conditions based on interference caused, being progressively introduced around the world in the next decade.

10.2.5 Harm claim thresholds

In the US an analogous approach has been proposed termed "harm claim thresholds" [2]. The authors propose that the FCC establish in-band and out-of-band interfering signals that must be exceeded before a system can claim that it is experiencing harmful interference. The aim is to remove any uncertainty in the rights that a license holder has to avoid interference and provide them with clear guidelines against which to design their receiver performance. The proposal calls on the regulator to modify spectrum licenses such that they define the signal levels a license holder needs to tolerate before they can bring a claim for harmful interference. Hence, if a license holder chooses to procure receivers with insufficient protection and then suffers interference it cannot claim protection from the regulator, as happens today.

The proposal would establish a "field strength profile"[1] due to neighboring signals that is defined both inside and outside the licensed frequencies. This threshold must be exceeded at more than a specified percentage of locations and times in a given area before the affected operator can bring a claim against the neighbor. Thresholds can be set according to neighboring use, expected future developments, and other parameters, and hence can vary from band to band and even from operator to operator.

At present, these are just proposals, but with the continual stream of problematic interference cases in the US there is a strong incentive to implement a solution of this sort.

10.3 Receiver standards

10.3.1 Introduction

The issue of receiver performance was raised in section 2.5.2 above, where it was noted that receiver performance was one of the key determinants of whether interference was experienced. Poorly performing receivers effectively reduce the efficiency with which spectrum can be used, but market forces tend to drive manufacturers toward poor receivers. Hence, there is some argument that regulators or others should set standards

---

[1] This would comprise the maximum allowed levels of radio signals as measured by their power and is equivalent to the interference measure defined for SURs.

for receiver performance and take appropriate action where these standards are not met. However, despite having discussed this point for decades, regulators are still to take action. Broadly, this is because (1) they do not have the necessary legislative tools and (2) it is far from clear what the most appropriate action would be. This section considers the issues and suggests some possible solutions.

10.3.2 Examples of issues

There have been many cases where poorly performing receivers have limited the use of spectrum or even prevented the introduction of new services. For example:

- The prospect of overload interference to legacy satellite digital audio radio service (SDARS, aka SiriusXM) receivers from mobile devices in the wireless communications service (WCS) required application of strict technical rules and effectively created 5 MHz guard bands on each side of the SDARS allocation.[2]
- Many C-band satellite earth station receivers operating at 3,700–4,200 MHz are susceptible to signals from well inside the 3,650–3,700 MHz band that was transferred from federal to commercial use, risking the possibility that much of the federal transferred spectrum would be useless [4].
- The use of the 20 MHz AWS-3 band (2,155–2,175 MHz) for time division duplex operation was blocked because cellular handsets in the lower adjacent AWS-1 F-block (2,145–2,155 MHz) were designed to operate across the AWS-3 spectrum consistent with international (but not US) allocations and thus were unable to reject interference from nearby AWS-3 handset transmissions [5].
- The AWS-1 downlink spectrum at 2,110–2,155 MHz is upper adjacent to the broadcast auxiliary service (BAS) at 2,025–2,110 MHz. AWS-1 licensees were required as the newcomers to correct any harmful interference to the BAS operations. Since BAS equipment had not been designed with sharp filters, AWS-1 operations were found to cause harmful interference to BAS, requiring the AWS-1 licensees to pay to design, purchase, and install new filters for BAS equipment.[3]
- TV receiver performance was a significant issue for the access of unlicensed devices to the TV white spaces. The roll-off of the TV filters is the dominant factor limiting the amount of energy that a TV white space device may emit in the white space and therefore the potential applications for the devices.
- In the UK, poorly performing radar receivers at 2.7 GHz delayed the introduction of mobile services in the 2.5–2.7 GHz band for about five years and cost the government many millions of pounds while new filters were retro-fitted to the radars [7].
- Receiver performance relative to adjacent channel and intermodulation characteristics was a major element in the issue of rebanding the 800 MHz spectrum to avoid interference between Nextel and public safety operations on interleaved channels [8].
- LightSquared's proposed deployment of ancillary terrestrial component (ATC) base stations as part of a hybrid terrestrial–satellite service has raised significant concerns

---

[2] A useful presentation on this topic is available at [3].
[3] A presentation from the FCC on this topic can be found at [6].

about potential harmful interference to the GPS service operating in the upper adjacent spectrum due to the potential for receiver overload, i.e. power transmitted in LightSquared's licensed frequencies causing degradation of GPS devices that did not filter out this energy sufficiently well.[4]

Most of these examples relate to the US – primarily because it is in the US that new services are typically first proposed and hence these issues tend to surface. It is clear that there are many cases where receiver performance has been a key issue and the trend appears to be for the number of cases to grow ever faster as more users are packed into the spectrum.

10.3.3 Interested parties

One of the problems with regulating receivers is that there are many interested parties, including

1. the various actual or potential providers of spectrum-using services (e.g. broadcasters),
2. manufacturers of reception equipment (e.g. TV manufacturers), and
3. end users – who normally purchase and own reception equipment (e.g. viewers).

Their interests will not be aligned and may not even be linked (e.g. broadcasters and viewers have little economic or regulatory linkage between them).

If we take a service such as digital TV or cellular handset reception, the supply chain is typically made up of a small number of large international suppliers (typically of the order of five suppliers making up more than 80% of the global market) and a larger number of license holders (typically two to five in each country, making many hundreds in total). Hence, the manufacturers often have more power in equipment design than the operators – even if an operator decided not to recommend a particular manufacturer this would only make a small difference to that manufacturer's market base.

In some cases, there may be international groupings of license holders; for example, Vodafone owns licenses in many countries. Nevertheless, any one license holder is typically only a relatively small customer of one or more manufacturers. Hence, the ability of a single license holder to enforce a request for a manufacturer to produce receivers with better specifications may be limited. Only if the license holders act collectively is it likely that the equipment manufacturers will respond.

Collective action is entirely possible, and there are entities such as the GSMA which enable this, but the need for collective action generally makes it harder to achieve a desired outcome.

In the case where receivers will actually fail there may be an added incentive for manufacturers to avoid bad publicity, but in the cases where license holders are seeking to negotiate more efficient interference levels between them then this incentive may be lacking.

---

[4] See FCC notice on suspension of the waiver that LightSquared had been operating under at [9].

There is a weak link in the world of TV where the broadcasters typically have no control over the purchasing choices of the viewers and limited ability to advise them on receiver quality. Viewers are rarely sufficiently well informed to make a purchasing decision on the basis of the filter in the receiver and its ability to cope with prospective changes of use in neighboring bands, and just assume that all TVs will work adequately well at receiving signals. In such a case, even if a license holder wished to address the problem, there might be limited means at its disposal to do so.

10.3.4 Too big to fail

There are some services where there are policy or societal reasons why failure is not palatable – TV reception is an obvious example. Much of the population would find it unacceptable to have to change their TV receiver as a result of a change of spectrum use in a neighboring band and political pressure would likely occur to find an alternative solution to the problem. Of course, if the manufacturers realize this then there is no incentive on them to improve receivers since they effectively have a guarantee that whatever level of performance they select, neighboring services will be tailored to fit with this.

This is the "too-big-to-fail" problem. It is the ability to use neighboring bands for alternative applications that suffers as a result. It means that it would not be rational for a farsighted end user to purchase a more expensive receiver, because it would never be necessary.

10.3.5 Options for regulating receivers

A number of options have been put forward for managing receiver performance:

1. leaving it to market forces
2. regulatory involvement "behind the scenes," and
3. direct regulatory action.

The first – market forces – has not worked to date and there is little reason to believe this will change. The PCAST report did propose that the scope of the auctions of shared access rights would include the externalities of the requested receiver protection, providing some degree of financial incentive to minimize receiver protection constraints on other users. Whether this could be implemented and whether it would have any real impact is far from clear at present.

The second involved the regulator using their influence behind the scenes, most likely in standards activity. Many standards will mandate the signal levels at which receivers should work – the minimum sensitivity level – and devices will be tested to see if they can operate with this signal. Many exceed the level by a good margin but this cannot be relied upon by regulators as not all will, and future devices may not. Some entities, such as the DVB group, set recommendations for receiver performance and issue "tick marks" to those TVs that meet these guidelines. However, these levels are not set for spectrum management purposes but to optimize commercial operation and hence may not be sufficiently good.

A better approach would require standards bodies to consider possible alternative uses in neighboring bands and, where appropriate (i.e. after conducting a cost–benefit analysis), to specify receivers capable of accommodating such uses. At present, standards bodies typically do not have the expertise to undertake such analysis, so it might be better accomplished by more regulatory involvement in the standards process. The key problem is that regulators have limited resources to attend myriad standards meetings and to provide considered input. It is also not a guaranteed solution – standards bodies will likely be sympathetic to regulatory input but they are under no obligation to adopt proposals from regulators. Nevertheless, the generation of better standards would seem to be one tool that is already available, and an area where regulators should play a larger role, ideally acting in concert.

Another area of behind-the-scenes activity might be to "name and shame" poor-quality receivers, thereby both giving the public a warning not to buy particular brands and encouraging manufacturers to improve. Whether this would be effective would depend on the ability of the regulator to disseminate technical information widely to consumers. However, it might risk manufacturers bringing litigation against the regulator if they believed the regulator was damaging their business, and so would have to be undertaken with great care. To date, regulators have seen this route as too risky, and possibly outside their mandate. Again, concerted action from regulators might overcome some of these problems and be sufficient for manufacturers to take real notice.

The last approach is direct action, such as informing license holders of the performance expected of their receivers and introducing new services in neighboring bands on this basis. Regulators tend to try to avoid this as it can impede the actions of the spectrum market, but in some cases it may be necessary. License conditions could be set on the basis of an assessment of realistically achievable requirements rather than the minimum required performance, or could be modified over time as receivers progressively improve (although it is not clear that this will always occur). However, as discussed above, unless manufacturers take note of such conditions and act accordingly, such an approach may not overcome the problems identified earlier. It also requires the regulator to take a view as to what can actually be achieved, which the regulator may not be well positioned to do.

Alternatively the regulator could take direct action in specifying receiver characteristics. This could require all receivers of a particular type sold in the country to meet a better specification than previously. There are many variants of this – for example, it might start by only applying to higher-cost devices operating on a longer replacement cycle, where the additional cost of the enhanced receiver is less noticeable, and then be extended progressively to other receivers as economies of scale improve. Such intervention should be preceded by a detailed assessment to show that the intervention was merited and proportionate.

10.3.6 Where should the limits be set?

A key assumption in all of these solutions is that it is possible to determine what the "right" level of receiver performance should be. However, this is far from simple. If the level is set too weak (so poorly performing receivers are allowed) it has little impact or may even encourage worse receivers to be developed. If it is set too strong (receivers must perform very well) it will impose unnecessary costs on consumers or those buying receivers. If it is set for too long a time period it might miss changes in technology that allow for better receivers. If it is set nationally it might result in manufacturers not supplying a relatively small market or demanding a large premium. But if it is set internationally it may be difficult to find a level that suits all and the studies and negotiations may take so long that they are out of date by the time they are concluded.

These are difficult problems. However, they are of the same level of challenge as other problems faced by regulators, such as the license conditions allowed. There is no simple answer but careful study by experts in the field should result in limits that are sufficiently close to optimal and certainly better than those used to date.

It is also worth noting that SURs provide a direct solution to the problem in the form of the indicative interference levels that are inherent in their design. However, this still requires care in setting the SUR interference levels and SURs are not yet implemented to any extent.

10.3.7 Relevant studies

There have been a number of pieces of work in this area. A study that TTP conducted for Ofcom [10] showed that 3–6 dB gains in receiver performance in terms of better filtering could be made for typically one or two dollars' additional device cost. For many devices, such as TV receivers, this would be a minimal cost that would lead to real improvements in spectrum efficiency.

A more recent TTP study – "80dB DTT ACS: Is It Possible, and at What Cost?" [11] – looked at ways to achieve 80 dB ACS (adjacent channel selectivity) in DTT receivers, which was thought to be sufficiently good that ACS would never be the dominant mechanism by which receivers suffered interference. TTP noted that the best-in-class receivers almost achieved this today, but that there are many poor receivers. While there are many ways to improve performance, cost pressures tend to drive toward lower-performing solutions. They noted that the current approach of voluntary testing against national standards is not delivering best-in-class performance and that mandatory standards are difficult to introduce, although this could be achieved at a European level. They suggested that self-imposed agreements could be more proportional but were unsure that they would work.

They concluded by saying, "Ideally the issue of DTT performance would be addressed through a relevant standard that was difficult for manufacturers to avoid and easy for consumers to recognise backed by legal force." But it is far from clear how to achieve this.

The EC have also looked into this issue. A precedent for a change in policy was set by the Radio Spectrum Policy Programme (RSPP). Article 3k of the RSPP says that the immunity of receivers to interference should be increased in order to boost the efficient use of spectrum.[5] In addition, in June 2013, the Radio Spectrum Policy Group (RSPG) published its "Opinion on Strategic Challenges Facing Europe in Addressing the Growing Spectrum Demand for Wireless Broadband" [12]. The document says that spectrum efficiency could be increased through tighter regulation of receivers.

Recommendation 4d of the opinion says there is "a need for a clear EU policy on improving spectrum efficiency, where it would be an essential requirement to construct TV receivers so as to avoid harmful interference." Similarly, in July 2013, the RSPG adopted a report on interference management [13] which, inter alia, looked at

- ways to promote improved receiver standards within the current ETSI and CENELEC processes and the EU institutional setup, as well as to indicate how the role of European institutions could facilitate such a breakthrough, and
- examination, through the analysis of best practices, of what role EU spectrum policy and specifically the R&TTE and EMC Directives could play in promoting improved receiver standards.

10.3.8 Summary

Regulators are well aware of the benefits of receiver standards but have not implemented them because of the difficulties in this area. However, the pressure is increasing as services are packed more closely together and as sharing moves up the agenda. Self-regulation, market forces and existing standards approaches do not seem sufficient, hence it seems likely that there will eventually be direct intervention to set standards. Quite what form this will take is unclear. We suggest that the lead needs to come from regulators in making it clear, ideally at a multinational level, what level of performance they expect from receivers in each frequency band. This might come initially in the US, where the national market is large enough to support a "go-it-alone" approach. This may be sufficient to persuade standards bodies and manufacturers, but if not then regulators need to be ready either to use legal tools in some way or to push ahead with licensing decisions that would be unpopular with politicians and others.

References

[1] See http://stakeholders.ofcom.org.uk/consultations/sur.
[2] J. Vries and P. J. Weisner, "Unlocking Spectrum Value through Improved Allocation, Assignment, and Adjudication of Spectrum Rights," the Hamilton Project, discussion paper, 2014-1.

---

[5] Decision No. 243/2012/EU of the European Parliament and of the Council of March 14, 2012, establishing a multi-annual radio spectrum policy program.

[3] See http://transition.fcc.gov/bureaus/oet/receiver-workshop1/Session4/SESSION-4-6-Schaubach-WCS.pdf.
[4] See Comsearch, "Estimating the Required Separation Distances to Avoid Interference from Part 90 3650–3700 MHz Band Transmitters into C-Band Earth Stations," at www.comsearch.com/files/TP-102516-EN_LR_3650-3700_MHz_Interference_into_CBand_ES.pdf.
[5] T-Mobile, "AWS-3 to AWS-1 Interference," at http://apps.fcc.gov/ecfs/document/view?id=6520035723.
[6] See http://transition.fcc.gov/bureaus/oet/receiver-workshop1/Session4/Session-4-4-NAB-Victor-Tawil.pdf.
[7] Ofcom, "Notice of Coordination Procedure Required under Spectrum Access Licences for the 2.6 GHz Band: Coordination with Aeronautical Radionavigation Radar in the 2.7 GHz Band" (March 2013).
[8] See http://transition.fcc.gov/pshs/public-safety-spectrum/800-MHz.
[9] See www.fcc.gov/document/spokesperson-statement-ntia-letter-lightsquared-and-gps.
[10] Ofcom, "Study of Current and Future Receiver Performance," at http://stakeholders.ofcom.org.uk/market-data-research/other/technology-research/research/spectrum-liberalisation/receiver.
[11] See http://stakeholders.ofcom.org.uk/binaries/spectrum/UHF700MHz/DTT_RX_study_stakeholder_presentation_20131125_released_20131122.pdf.
[12] See RSPG13–511 Rev 1.
[13] See RSPG13–527 Rev 1 final.

Part IV

Case studies and conclusions

11 The struggle for the UHF band

11.1 The issues at stake

Territorial wars are seldom waged over infertile land, unless it has military or symbolic significance. The same is true of spectrum. Given the technologies available today, the most fertile spectrum lies below 3 GHz, and the very best of it is found in or just above the UHF TV band, which falls between 400 and 800 MHz.

This spectrum is particularly well suited both to terrestrial broadcasting and to mobile communications as a result of its propagation characteristics. At these frequencies radio signals travel many tens of kilometers, but equally receivers do not require antennas that are larger than handheld devices. In the case of broadcasting this allows most of a country to be served from a relatively small number of high-tower sites (e.g. around 100 for the entire UK). In the case of mobile communications the extended range is valuable both for covering rural areas and for achieving better in-building penetration than could be achieved at other frequencies.

In both cases, there are other options that can be used to enable service delivery. Broadcasting can be delivered using other spectrum bands or alternatively via non-spectrum inputs such as fibre-optic cables. TV distribution via satellite broadcasting uses spectrum in the 12 GHz range and this is much less valuable than the UHF band. Coaxial and fibre cables can be used to distribute TV content, which is increasingly the case in dense urban settings.

With regard to mobile communications there are many higher bands available – notably at 1,800 MHz, 2.1 GHz, 2.6 GHz and others. Voice and data services are also available for fixed and nomadic consumption by wireline technologies, combined with Wi-Fi final drop. But auction data show that the higher frequencies are much less valuable (see Chapter 5). Telecommunications regulators continue to regard (perhaps with decreasing conviction) fixed and mobile voice and data services as falling in different economic markets, because they are seen by users as being insufficiently substitutable for one to impose a constraint on the other.

This may change. Current discussions of spectrum for 5G mobile, expected to be available sometime between 2020 and 2025, often revolve around use of much higher bands – as high as 60 GHz. With an appropriate change in the installed base of TV sets, a shift entirely away from terrestrial transmission to cable or satellite is imaginable. Broadcasting and mobile communications might converge on the same transmission technology. But for the moment, the competition between the two for the UHF band is intense.

This situation highlights the importance of interference issues. As explained below, terrestrial television is transmitted on a high-tower, high-power basis. Transmitters can interfere with one another in the same and in adjacent bands. For this reason, spectrum assignments have to be planned on an international basis, and a major national change, such as the introduction of digital terrestrial broadcasting, has to be carefully co-ordinated. This introduces a strong element of central direction into spectrum assignments for broadcasting, which also spills over into the management of contiguous mobile spectrum.

From the standpoint of spectrum management, rivalry for the UHF band is of interest for another reason. It might appear that there is a quick and easy way of resolving the conflict: why not simply hold an auction or series of auctions for the disputed spectrum? The amount that a firm in either sector can bid is ultimately determined by the willingness of its customers to pay for the service provided. On this basis, the firms winning the auction would have behind them the customers willing to pay the most for the relevant service to be provided.

The problem here is that the private value derived directly by customers of the service is not the only benefit. For example, it is claimed of mobile voice and data services that their value to the economy is not captured entirely by benefits which accrue to the direct customers of the services. For example, online customers for a product may make a sector more competitive and benefit all customers. This argument is often supported by claims that a 10% increase in the penetration of both voice and data services (for example from 30 to 40%) increases GDP permanently by up to 1.5%,[1] with benefits accruing to others than their direct purchasers or people with an immediate monetary relationship with their direct purchasers.

Similar claims, with a longer history, have been made in relation to broadcasting. It is claimed for public-service broadcasting in particular that it generates an educated and cultivated citizenry, which is a broader social benefit to all; that it combats exclusion; that it brings a nation together; and so on.

These issues have been widely debated in relation to the allocation of the UHF band. The band thus provides a topical case study in spectrum management, illustrating a range of issues. In this chapter, we first set out the claims for the UHF spectrum of broadcasting and mobile communications separately, and then discuss how they might be resolved.

11.2   Broadcasting, the digital switch-over, and current trends

Broadcasting of both radio and television services relies on an increasing range of delivery mechanisms, including terrestrial broadcasting, satellite broadcasting, and Internet delivery using non-spectrum-based cable, copper, or fibre networks. In this chapter we are primarily concerned with television broadcasting since it is a much greater user of the spectrum than radio broadcasting (around 20 times more spectrum).

---

[1] See section 13.3 below.

Further, our concern is with terrestrial broadcasting, since satellite broadcasting has few pressures for the spectrum and Internet delivery is broadly via wired solutions that are outside the scope of this book. Nevertheless, the fact that these alternatives are present is critical for the discussion that follows in that it enables, in principle, a switch from terrestrial broadcasting if users can be persuaded to move to alternatives.

Television broadcasting has a history extending back some 90 years.[2] During that time much has changed, but the idea of using a few very high masts to cover large areas remains. A typical broadcasting tower might be 100–200 m tall and use a transmitter power of 100 kW or more. This compares to a cellphone tower height of 10–20 m and power of less than 100 W. For this reason, broadcasting is sometimes described as "high-tower, high-power" (HTHP). This enables a country like the UK to be mostly covered with around 100 tower sites, compared to the 10,000+ needed for equivalent cellular coverage. These 100 HTHP masts are supplemented with around 10 times as many relay stations, which are lower-power transmitters used to fill in particular coverage holes such as deep valleys.[3] However, coverage is typically only planned to rooftop directional antennas. This means that indoor coverage is unreliable and mobile coverage may only work when relatively close to the transmitter. Broadcast networks are also only one-way, sending information from the transmitter to TV sets but not accepting information in the opposite direction.

With the low frequencies used and the high towers, TV signals travel a long way. Successful reception of TV signals is often possible 50 km from the transmitter, and some signal energy can still be present over 100 km away, potentially causing interference if the frequency is reused. This means that for many countries TV transmitter frequencies need to be agreed on an international basis – for example in Europe agreements cover something like 43 countries stretching from Ireland to Russia.[4] It also means that spectrum usage can be inefficient because adjacent transmitter zones cannot use the same frequencies. The map-colouring theorem tells us that this needs four different frequencies for each transmitted channel across an area, and in practice more than this are sometimes used.[5]

These factors tend to make broadcasting very hard to change. Any change of frequencies requires the agreement of tens of countries or more in a costly process that takes years to conclude. Any movement of transmitters requires homes to repoint their antennas. Even changing frequencies on the same transmitter may require a different aerial in the home as TV aerials tend to be different for the lower and upper parts of the UHF TV band, As a result, viewers will need to rescan all their receiving equipment.

Where TV has changed over the years is in the format of the delivered signal. Initially this was black-and-white, with 405-line definition. Then, in 1954 in the United States and gaining momentum in the 1960s and 1970s, broadcasters around the world changed

---

[2] The British Broadcasting Company, jointly owned by radio set manufacturers, began operations in 1923. It became the publicly owned British Broadcasting Corporation in 1926.
[3] Overall, the UK terrestrial transmission network reaches 98.5% of homes.
[4] The latest is termed GE06 because it was concluded in Geneva in 2006. Details can be found at [1].
[5] The well-known theorem states that for any arrangement of shapes drawn on a page it is possible, using four colours, to colour them such that no bordering shapes have the same colour.

to colour TV with improved definition (around 625 lines outside the US and Japan). The next major change, still in progress, was from analogue to digital. Digital transmission of video information is much more efficient than analogue. This is because a TV picture comprises a sequence of still pictures, typically sent 25 times a second. Successive pictures are nearly identical; this is only not the case when a change of scene occurs. Analogue transmission sends each still picture regardless, whereas digital systems encode the difference between each picture rather than the whole picture. This can often be a tenth or less as much information. Hence, when digital delivery was introduced it was possible to fit some six to eight times as many channels into the same bandwidth as one analogue channel. This enabled three gains:

- more channels/choice, with some countries moving from around four to nearer 40 channels;
- higher-definition broadcasting (see below); and
- some spectrum being returned for other uses – this is often called the "digital dividend."

TV definition has also improved. High-definition transmission has approximately twice as many lines as standard definition (1,080 lines is an often-used format) and even higher-definition systems with as many as 4,000 lines are now emerging, such as UHDTV.[6] However, higher definition requires more information to be sent – some two to three times as much as for standard definition, reducing the number of channels available, especially if channels need to be transmitted in both standard and high definition (this process is termed "simulcasting") so that viewers without high-definition receivers can still view the content. Three-dimensional formats have also been trialed (3D-TV) but to date have not proven popular and have broadly been discontinued by broadcasters.

Mobile broadcasting (reception of TV on mobile devices) has been trialed repeatedly over the years, with portable units introduced in the 1980s and then mobile-phone-based reception trialed in the 2000s. Mobile TV solutions are typically very similar to standard-broadcast TV but with changes that reduce the power requirements on handheld devices. Systems of this type have been deployed around the globe but to date have all failed (except where they have heavy government subsidy in South Korea). This appears to be for a range of reasons, including that

- reception is patchy for handheld devices at street level;
- extra equipment is needed in the handset that manufacturers do not generally wish to include;
- the business case is unclear; and
- users generally do not want to "tune in" when mobile, but instead to watch stored podcasts or similar.

---

[6] In June 2014, the FIFA World Cup of that year (held in Brazil) became the first to be shot entirely in Ultra HD, by Sony. The European Broadcasting Union (EBU) broadcast matches of the FIFA World Cup to audiences in North America, Latin America, Europe and Asia in Ultra HD via SES' NSS-7 and SES-6 satellites.

Some still believe that mobile broadcasting will eventually find a role, especially with the proliferation of tablets and larger-format smartphones.

11.2.1 The digital switch-over

The twin logics of the switch from analogue to digital broadcasting transmission are spectrum efficiency and improved picture quality. However, the main spectrum dividend only comes when analogue terrestrial transmission is switched off, and this can only be accomplished after a period of "simulcasting" or simultaneous transmission of terrestrial signals in analogue and digital form, in the course of which all or very nearly all households equip themselves with a new television set or set-top box capable of receiving digital signals. Digital reception allows the viewer access to many more channels than analogue. But in many countries, the process has been completed with the assistance of set-top-box subsidies for late adopters or households satisfying certain criteria. The prospect of releasing and, in most cases, auctioning the spectrum released by analogue switch-off has encouraged governments to go down this road.

Digital transmission uses a number of different standards around the world, one of the most popular of which is digital video broadcasting – terrestrial (DVB-T), used in Europe and elsewhere. DVB systems multiplex a number of TV channels onto a single radio channel which is 8 MHz wide (the same size as the old analogue transmissions). The number of channels in the multiplex depends on many factors, including the desired transmission quality, such as standard or high definition, and the required robustness of the received signal, where the multiplex capacity is reduced when longer range is needed. Typically around five channels might be transmitted on the same multiplex. More recently, an enhanced version known as DVB-T2 has been introduced which increases the multiplex capacity. Coupled with more advanced video codecs this allows up to twice as many channels per multiplex. However, new receiving equipment is needed.

11.2.2 Broadcasting trends and their impact on spectrum management

In the course of digital switch-over and after analogue switch-off, legacy analogue networks have generally kept their prime positions in free-to-air television viewing. However, three further trends are visible. First, the share of pay-TV viewing grows, where an over-the-air signal has to be encrypted. Second, there is a gradual erosion of the combined market shares of the legacy analogue channels, as viewers gain access to a wider choice of free-to-air channels. Third, while the viewing of traditional "linear" channels, in which programs are broadcast in accordance with a pre-announced schedule, has fallen, but not too precipitately, younger viewers in particular are less attracted to such consumption, and rely increasingly on "nonlinear" on-demand video offerings, of a kind often available on social media and accessed by nonconventional broadcasting delivery networks, often to smartphones or tablets.

As these trends gather strength, it is likely that broadcasting's reliance on UHF delivery will progressively diminish. Already in many jurisdictions spectrum managers have refarmed some UHF broadcasting spectrum to mobile communications purposes. The pressure to do more of the same is likely to grow. The next section discusses some future options for broadcasting transmission.

11.3 Broadcasting technical options

11.3.1 Introduction

In considering the future of the UHF band there are many technical options that could be pursued. These can broadly be divided into:

- using less spectrum to continue broadcasting,
- combining broadcast and cellular networks in some manner, and
- stopping all UHF TV broadcasting in favor of alternative delivery mechanisms.

Here we consider the first two.

11.3.2 Using less spectrum

The amount of spectrum needed to broadcast a given level of TV content (a given number of channels at a given definition) is related to (1) the amount of spectrum per channel and (2) the amount of reuse of spectrum needed to avoid interference in neighboring areas. The amount of spectrum per channel is dependent on how much the content can be compressed prior to transmission and the efficiency of the transmission system. Broadly, no further gains are expected in the transmission system, which is already close to theoretical limits on what can be achieved for the current configuration of transmitter sites. However, further gains are possible in compression. For example, Table 11-1 shows how compression has improved over time, resulting in ever less bandwidth being needed to transmit the same content.

If this trend were to continue it is possible we might see something like a halving in the bandwidth required over the next decade. This would either allow for half the spectrum to be used, or for more channels to be broadcast in high definition, or some combination of the two of these. However, the figures show that improvements tend to occur when a new version of the compression algorithm is introduced (e.g. MPEG-2, MPEG-4). This

Table 11-1. Trends in video coding

Standard	Date	Mbps/SD channel
MPEG-1/H.261	1993	35
MPEG-2/H.262	1995	25
MPEG-4/H.264	2004	12
HEVC/H.265	2013	4

results in a few years of significant gain, which then tails off until the next new algorithm. Almost invariably, a new algorithm requires new receiving equipment since typically it is much more processing-intensive than its predecessor, and old equipment tends not to have enough memory or processing power. New equipment is a major problem, requiring viewers to replace their TVs and ancillary equipment such as personal video recorders (PVRs). This takes many years to happen and, typically, during this period transmissions mostly occur in both the old and the new formats (simulcasting). This is less efficient than the original compression mechanism, so unless broadcasters have some "spare" spectrum that they can borrow during this period it can be very problematic to make the change.

The second approach is to reduce the amount of reuse needed. Recall that typically something like four times as much spectrum as is required in one cell is needed because neighboring cells need to use different frequencies. This is referred to as a multifrequency network (MFN) and is the approach used widely to date. TV standards do allow for all transmitters to use the same frequency – so-called single-frequency network (SFN). They do this by transmitting exactly the same content at exactly the same time. After each "bit" is transmitted a gap is left so that signals from more distant transmitters can arrive. Then the next bit can be transmitted. The gaps add a little inefficiency (some 10–15%) but this is small compared to the savings of not needing to reuse frequencies. The problem with SFNs is that the same content needs to be broadcast from all transmitters. This is not possible at national boundaries since each country has different content. Even within a country there is often regional variation both in the content and in the advertisements. Within the constraints of areas like Europe, where some countries have more than ten neighbors, there is little advantage in using SFNs since they would only have an impact in the center of the country at best. SFNs are also not viable in the US, where different content tends to be broadcast in different cities. So while there is much promise from SFNs, it seems unlikely that they will be introduced in most countries.

So, in summary, short of reducing the number of channels, the scope for significant savings in the amount of spectrum used for DTT is limited and may take decades to realize.

11.3.3 Combining platforms

A different approach is to seek to consolidate broadcasting and cellular onto the same "platform." This might allow efficiencies in transmission as well as facilitate convergence between the two services and potentially improve the prospects for mobile broadcasting.

For some years cellular systems have sought to have a broadcast mode. The original intent was that if many users in a cell were streaming the same content then instead of delivering a separate stream to each, they could all "listen in" to the same broadcast channel. This would significantly reduce the loading on the cell, saving money for the operator. While an approach was standardized within 3G, it required spectrum to be set aside, and most operators did not have sufficient spectrum to do this, so it found little favor. However, a better solution was developed as part of the 4G/LTE specifications. Termed "evolved multimedia broadcast multicast service" (eMBMS [2]), it provides much more flexibility to dynamically allocate resource to a broadcast channel within a cell. At the time of writing it was in early trials in Australia.

Its availability has led some to ask whether all broadcasting could be moved from the DTT platform (by which we mean the network of transmitter sites and the DVB technology) to that of the mobile operators.[7] In this case the existing HTHP sites could be turned off (although they might be retained for radio broadcasting) and TV would be transmitted from the mobile phone low-tower, low-power (LTLP) sites. The key advantages of this might be

- lower operational costs;
- better integration between broadcasting and broadband, enabling interaction, easy switching between broadcasting viewing and streaming, etc.; and
- better spectrum efficiency.

We will focus here only on the last of these – suffice to say that the first two are far from proven. Might an LTLP approach provide greater spectrum efficiency? Broadly LTE is around as efficient technically as DVB-T2, the current technology preferred for DTT. There are some exceptions to this, for larger LTE cells, that we will return to. The key advantage of LTLP is in the ability to use SFNs. The main issues for SFNs occur at national boundaries. Simplistically, a cell will generate interference out to about twice its radius. So for an HTHP cell with a 50 km radius, the area where the frequency cannot be reused could extend 50 km into the neighboring country. But for an LTLP cell with a 5 km radius this is much reduced. This means that an SFN could be operated not just in the center of a country but in the vast majority of its area. Potentially this could reduce the need for spectrum by as much as a factor of four – from the typical 320 MHz in use today down to around 80 MHz.

However, the efficiency of the LTE eMBMS system depends on the cell radius [4]. This is because an eMBMS broadcast needs to be in a format that all in the cell, even those at the cell edge, can receive. As the cell gets larger, so the signal quality at the cell edge reduces, requiring low-order modulation and greater error correction. Roughly, beyond around a 5 km inter-site distance (ISD) – which means a cell radius of around 2.5 km, the efficiency of eMBMS falls below that of DTT, and by a 10 km ISD it may be as low as one-sixth to one-tenth of that of DTT, resulting in a lower overall spectrum efficiency, even if SFNs are possible. Given that part of the purpose of the mobile operators wanting UHF spectrum is to enable larger cells to improve economics of provision, a 10 km ISD might be standard, particularly outside urban areas.

Switching to eMBMS would be highly disruptive to viewers. They would need new receiving equipment and would probably have to reorient their rooftop antenna toward

---

[7] See EU Press Release on study in this area at [3]. In it, the commissioner, Kroes, said, "The TV viewing habits of young people bear no resemblance to that of my generation. The rules need to catch-up in a way that delivers more and better television and more and better broadband. Current spectrum assignments won't support consumer habits of the future – based on huge amounts of audiovisual consumption through broadband and IPTV."

the nearest mobile-phone mast. There would be no obvious incentive for them to do this. A different approach would be for the cellular operators to transmit DVB-T instead of LTE eMBMS from their towers. This would alleviate the need for new viewing equipment (although not for antenna reorientation) but would limit the possibilities for convergence.

All of these options would also need new commercial arrangements. It would not make sense for each mobile operator in a country to broadcast the entire set of TV channels. Instead, only one might be expected to do so. They would need to offer this service to all subscribers across all operators, raising competition issues and requiring guarantees that they would at least carry the public-service broadcasting (PSB) channels.

11.3.4 Conclusions

There are many technical options for the delivery of TV content and indeed for the future use of the TV bands. However, the transitional issues of moving from the current platform and viewing equipment are formidable and there are myriad international co-ordination problems. This suggests that progress will be slow – and there is a real risk that it will be overtaken by other trends, or a move away from a platform that is seen as out of date.

11.4 Mobile data, national broadband plans, and spectrum management

We now switch to the principal claimant to use of the UHF band – mobile broadband. Most countries have extensive ambitions relating to the diffusion of broadband, in many cases embodied in national broadband plans. This policy is championed by the Broadband Commission, set up by the ITU and UNESCO, which argues that the effective use of broadband networks, services, and applications can provide transformative solutions to address key issues of our times, including eradicating poverty and malnutrition, attaining healthy living for all, and decoupling economic growth from the use and depletion of natural resources [5]. The Commission states that to achieve these ambitious goals, broadband and information and communication technologies (ICT) must reach all people, in particular those facing social exclusion, living in remote locations, or facing highest vulnerability to environmental and economic factors. The key feature of broadband is that it is a general-purpose technology which influences almost all aspects of production and consumption, including the supply of social services such as health and education.

Many developing countries have no or very limited fixed networks, so have no choice but to rely on mobile (or, more generally, wireless) broadband. However, in developed countries broadband use is increasingly slanted towards mobile, with the growth of use of smartphones and tablets. Data on the diffusion of fixed and mobile broadband are regularly compiled by the European Commission (for members of the European Union), by the OECD (for its members), and by the Broadband Commission (for all countries).

In general terms the attainment of these ambitious goals will require primarily private capital, supplemented by public capital in poorer and more remote areas where private investment is not commercially viable. Experience with the worldwide diffusion of mobile voice suggests that such private capital can be forthcoming, but requires an environment which minimizes country risk arising from such factors as general economic instability and lack of security, and regulatory risk arising from unpredictable interventions and lack of regulatory consistency.

As far as spectrum management is concerned, this goal of reducing risk is promoted by such measures as clarity about the terms of spectrum licenses and about arrangements on the expiry of licenses, by the publication by the NRA of a spectrum strategy and a timetable for future spectrum awards, by appropriately chosen auction rules, by proper enforcement of interference rules, and so on.

The government may choose to give up some of its auction revenues to extend the coverage of the connected area. This can be done by making it a license condition that the homes of at least 80%, say, of a specified area are covered by voice or data services. If the constraint is binding, it is likely that auction revenues will be reduced. Clearly, coverage requirements may not be uniform for all countries or even across different regions in the country. They should meet the situation and needs of the country, and its ability to sacrifice revenue. Governments can simply impose an obligation to roll out the network to rural areas first (as in the case of Germany, for example, in the auction for the 800 MHz licenses), or they can be more specific, as in the case of Sweden, where the coverage obligation in one license in the 800 MHz auction was to provide service of at least 1 Mbit per second or better to a list of stated addresses identified as being broadband "not-spots," lacking any other forms of broadband connection [6]. A decision has to be made about whether to attach the coverage requirement to one license only, where it might create a monopoly in noncommercial areas requiring some form of price control, or whether to attach it to several.

There is a range of issues that arise in designing, monitoring, and enforcing coverage conditions. The government or regulator has to be sure that the investment and operating expenditures to provide the specified coverage are forthcoming; the need for certainty requires them to monitor the operator's coverage and to penalize the operator if it fails to provide service over it. This is important, since the government is offering the spectrum at a lower price in return for an expanded coverage. Thus there must be a credible plan in place to deal with non-performance. This plan must take into account the possibility that an operator may make a bid based on the expectation that if it gets into financial difficulties it will not be held to account for the extra coverage. It may be necessary to have in reserve a remedy in the license which allows the regulator to require the licensee to divest itself of the license by selling it to a third party, possibly together with any collateral investment wanted by a potential purchaser.

Another key requirement for the achievement of broadband plans is that available spectrum be awarded to licensees or assigned as a commons. Governments are sometimes tempted by large auction revenues to create an artificial shortage of spectrum to raise its price. This policy is usually short-sighted, since the effect of such artificial scarcity is to drive up service prices, to reduce take-up, and to curtail the benefits of broadband to raise income and, with it, the tax capacity of the economy. However, according to the view of spectrum auctions put forward in Chapter 3, using an auction to appropriate the scarcity rents associated with a band – for example the cost savings which an operator can achieve by acquiring 700 or 800 MHz spectrum compared with those available with 2.6 GHz spectrum – does not carry with it the expectation of higher service prices; all it does is transfer some of the scarcity rents from the operator awarded the license to the government.

11.5 Smartphones and the data crunch

The "mobile data explosion" forms an interesting case study for spectrum management – and one that is still unfolding. Usage of cellphone systems had only been changing slowly over the years, enabling periodic auctions to more than keep up with demand. The introduction of 3G and the high prices paid in the auctions around the year 2000 were based on assumptions by the mobile operators that data usage would grow dramatically, but for most of the 2000s there appeared to be little interest from the public in this. This all changed on January 9, 2007, when Steve Jobs introduced the iPhone. The ease of use of the touch screen led to very dramatic increases in mobile data use, not helped by the iPhone being a rather profligate user of network resources. Within the space of a few months networks were experiencing congestion, and within a year it was clear that there was a massive increase in demand under way.

Figure 11-1, looking at a specific four-year period from 2007 to 2011, shows how from almost nothing in 2007, mobile data traffic grew to the same as voice traffic in just under three years, doubling overall network load, and then within the next year almost doubled again. This is predicted to continue, as shown in Figure 11-2 (note: an exabyte = 1,000 petabytes).

Cisco and others [7] are predicting a growth of 66% each year out to 2017, resulting in 13 times more traffic in 2017 than in 2012, 100 times more than in 2009 and 1,000 times more than in 2008.

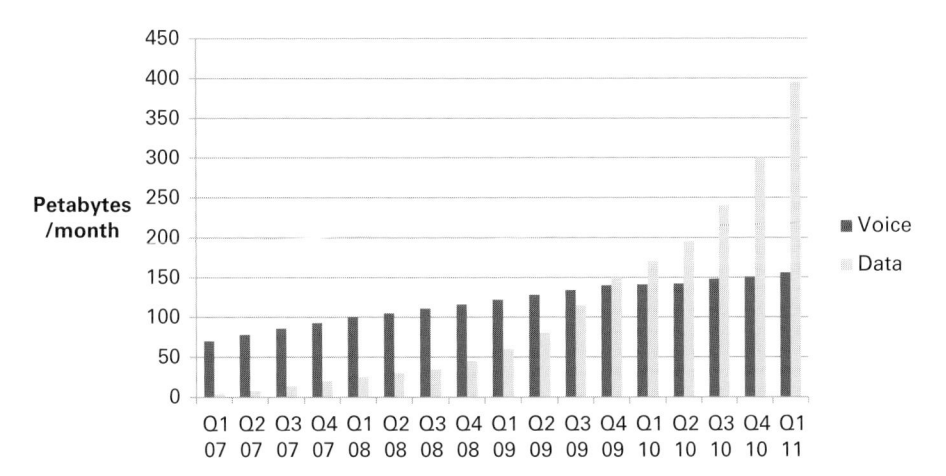

Figure 11-1. Global mobile voice and data traffic 2007–2011. Source: Web-data.

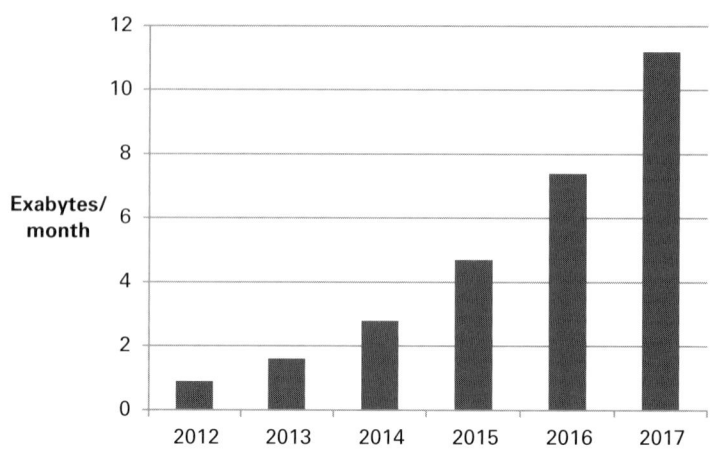

Figure 11-2. Predictions of global mobile traffic growth. Source: Web-data.

This massive growth left the operators struggling to meet demand, or at the very least to choke it off via price rises or rationing. Broadly, there are three ways to increase network capacity:

- greater technical efficiency,
- more spectrum, and
- smaller cells.

At the time, improved technical efficiency was not possible without introducing a new technology (as has subsequently happened with 4G). Operators put in place additional small cells as rapidly as possible, but also sought help from regulators around the world with urgent calls for additional spectrum.

Governments, if not regulators, were receptive to this, seeing it as a way to improve broadband penetration, stimulate innovation, and show they were facilitating a very popular service. This led to calls from presidents and similar for large increases in the amount of spectrum available for mobile broadband. For example, the US President [8] has called for the provision of 500 MHz of additional spectrum by 2020. Similar calls have been made in other countries. In finding this spectrum, policies that encourage innovation have not been at the forefront of regulatory thinking – simply responding to the needs of previous innovation has been sufficient.

Finding and repurposing spectrum is always problematic and becoming ever more so as spectrum is increasingly congested. Achieving the presidential mandate has been very challenging for all. In a 2012 speech [9], the chairman of the FCC set out how the FCC had met a mid-term target of liberating 300 MHz by 2015 through a mix of policies and spectrum bands, including:

1. auctioning 75 MHz of AWS spectrum in 2013;
2. removing restrictions on use of 70 MHz of spectrum, including 40 MHz of mobile satellite spectrum and 30 MHz in the WCS band;

3. using incentive auctions to free up more spectrum in the UHF bands, although the amount here is unclear – perhaps 100 MHz; and
4. using spectrum sharing, predominantly TV white space, to provide additional spectrum access – amount hard to define.

Whether this really does liberate 300 MHz of spectrum is open to dispute. Many of the bands discussed are being "modified" rather than "liberated" and outcomes, especially of the last two points, are far from certain. Most of these bands are still not yet in use – it takes many years for spectrum to be assigned, equipment to be manufactured, and handsets to make their way into a large percentage of the population. So, in practice, mobile operators had to overcome the data explosion with a mix of small cells and data caps until they gained additional spectrum in the 4G auctions.

In fact, much of this can be seen as posturing from all involved. Governments could argue that they set tough targets and stimulated action, while regulators could argue that they moved with alacrity to meet an important need. But in practice very little changed beyond plans already set in place. However, it has led to a different mindset – one where it is now assumed that mobile operators will have an unquenchable appetite for spectrum and that whatever spectrum can be found will immediately be devoured by them, sometimes to the detriment of other users. And, as we saw in the previous chapter, it has stimulated debate around the long-term future of the UHF band on the assumption that there is a pressing need for the spectrum from mobile operators.

Of course, there is another response to this problem, which is rarely discussed. Operators could simply increase the price for data access until demand and supply are aligned. However, most seem to believe that user demands must be met at a fixed price level however fast they grow. It is hard to see how this can make commercial sense and it is more likely that data costs will rise over time, especially for high-usage applications.

11.6 Resolving noneconomic valuation issues

Before addressing the UHF dilemma directly, it is worth discussing further one of the points raised above, concerning the valuation of external or indirect effects of spectrum use. An efficient system of spectrum management involves seeking an allocation such that, at the margin, the benefits of spectrum in the same band are the equal in use A and use B. Those marginal benefits include both private benefits directly accruing to users of the services competing for the input, and external benefits, which may primarily be economic but may also be social or political. They may include an improvement in general economic efficiency made possible through the more competitive markets emerging from enhanced connectivity in a country, or greater social solidarity achieved as a result of public-service broadcasting.

These external benefits arise not from use of the spectrum but from the provision of the relevant service. It flows from this that broadcasting or broadband (together with many more services of all types – for example, education or public transport) may be candidates for some sort of subsidy, in the same fashion that, in the case of services which generate an external harm, there is an argument for an extra charge on the service produced.

But getting from this proposition to a valid argument to subsidize the spectrum used to produce the service requires a series of intermediate steps. Giving privileged access to spectrum to broadcasters or mobile-communications firms at first sight looks like a bad way to subsidize a desired output, because it will encourage wastefulness in use of the input.[8] It is clearly much better to subsidize the output which generates the benefits than to subsidize one of the inputs into its production.

Second, when an external benefit to a service is imputed to an input for subsidy purposes, it is not clear how it should be done. For example, suppose that public-service broadcasting is found to yield an external benefit of $1 billion per year. A decision has to be made whether in the future to rely on terrestrial or satellite broadcasting as a delivery mechanism – a decision which depends on the rival claims of other potential users of the relevant spectrum. In arbitrating among those rival uses for the UHF band, the spectrum regulator would have to decide whether the external benefit of broadcasting should be imputed to that band, or to some other spectrum input, or to a non-spectrum input. Yet it cannot answer that question until it knows the efficient allocation, which is exactly what it is trying to find.

Now suppose that this hurdle has been overcome, and also that an estimate of the standard direct economic benefit of the service to customers has been estimated from willingness-to-pay data. How can we value the external benefit, bearing in mind, first, that ideally it will be valued in a way which makes it commensurate with the monetary valuation of the direct benefit, and, second, that we need to avoid double counting?

An exercise of this kind was, in fact, undertaken by Ofcom in 2009, as part of its Digital Dividend Review [11]. It was addressed to externalities associated with broadcasting.

The aspects of "broader social value" which were identified were [12]:

1. access and inclusion,
2. quality of life,
3. educated citizens,
4. informed democracy,
5. cultural understanding, and
6. belonging to a community.

They clearly present major measurement difficulties. These benefits typically are on top of the private benefits of the service to its direct consumer, and accrue to people

---

[8] This principle, which can crudely be expressed as "don't mess with input prices," has been shown rigorously in the case of taxation of inputs by P. Diamond and J. Mirrless [10]. To appreciate the operation of this principle, imagine that we chose to subsidize mobile communications by giving mobile operators free electricity. It might keep mobile prices down, but electricity consumption would certainly rise, and productive efficiency would be sacrificed.

other than that consumer; in other words, they are external effects. What would ideally be found is evidence of consumers' valuation of these benefits from observed behavior or revealed preference, but that is impossible to observe when such things as fellow citizens' levels of "cultural understanding" (one of the benefits for which a value is sought) cannot be bought directly in any market. An alternative method is to use a technique called stated preference, which involves asking a sample of people what value they place on the benefit which the inquirer describes. Another approach is to adopt a deliberative approach, for example by convening a "citizen's jury," which receives information about options and reaches a reasoned conclusion.

In the Digital Dividend Review, Ofcom considered the evidence on whether different levels of broader social value associated with different uses of the spectrum would be likely to result in a different ranking of alternatives with respect to total value, and appears to have concluded that this was unlikely to be the case.

The UK government in its 2014 Spectrum Strategy took this process further by committing itself to investigating this aspect of spectrum management policy more fully:

2.21. ... Spectrum value may be defined and measured in a number of ways. For example, in terms of final use (e.g. saving life) or cultural potentialities (e.g. social resilience or connectivity) or moral or cultural imperatives (e.g. the state has precedence over its use). In some cases, measures of these kind already exist ... Pending the development of more precise measures it may be useful in considering social or more intrinsic or hard to measure impacts to employ impact upon well-being measures (e.g. life satisfaction) as the valuation criterion ...

...

2.23. Planning the use of spectrum involves ranking the relative public values of various uses ... We will need to find ways to weigh social, economic, financial, technological and political factors against each other ... [13]

In the next section we assume that a valuation in money terms is available for such variables, and show by way of a stylized example how such valuations could be used to generate an efficient allocation of spectrum.

11.7 Finding an efficient allocation for the 700 MHz band

This section outlines one possible way of resolving a highly stylized conflict in the allocation of the 700 MHz spectrum between the demands of broadcasting and of mobile services. Although it is not intended to be realistic, the different scenarios considered do yield conclusions which differ in interesting ways from one another.

We start by considering the scenario in which there is a suitable alternative to 700 MHz spectrum for both mobile and broadcasting services. In practice, this will generally be the case. Then we discuss how the analysis is affected if either

a. there is no technically suitable alternative spectrum for broadcasting, or
b. the broadcasting service is not financially viable at alternative wavelengths.

Finally, we discuss the circumstances in which an auction-based system for assigning spectrum can be expected to result in an efficient outcome.

11.7.1 The illustrative scenario

In the first scenario, we suppose that (incremental) mobile services can be provided using either the 700 MHz or the 2.3 GHz bands. Similarly, broadcasting services can be provided terrestrially using the 700 MHz band or by satellite using the 10 GHz band. For simplicity, we assume that both the 2.3 GHz and the 10 GHz bands are otherwise unused.[9] This implies that there are two alternative assignments, as follows:

- Assignment 1: mobile services use the 700 MHz band and broadcasting the 10 GHz band.
- Assignment 2: broadcasting services use the 700 MHz band and mobile the 2.3 GHz band.

For the purposes of this scenario, we initially assume that

- both services are financially viable without subsidy in both assignments;
- output of both services is unaffected by the assignment of spectrum, so our focus is on the efficient production of a constant level of output.

Each service generates benefits to users and external benefits to nonusers, and results in a cost of supply. The value to society from each service is given by the sum of the user benefit (consumer surplus) and the external benefit to nonusers, less the cost of supplying the service.[10] Tables 11-1 and 11-3 show the assumed cost, user benefit, external benefit, and value to society for each service in Assignments 1 and 2. As can be seen,

- costs for both services are lower with 700 MHz spectrum than with the next-best alternative band (i.e. the 700 MHz band allows output to be provided at lower network cost);
- as output is constant, the user benefit from each service is the same in both assignments, and is higher for mobile than for broadcasting; and
- the external benefit from each service is the same in both assignments, and is higher for broadcasting than for mobile.

With these assumptions the broadcasting service would generate more value to society if it uses 700 MHz spectrum than would the mobile service, by virtue of the fact that it creates very significant external benefits. It would, however, be efficient to assign the 700 MHz band to mobile and the 10 GHz band to broadcasting, since the aggregate value to society from both services is higher in Assignment 1 (i.e. 1,080) than in

---

[9] If the alternative spectrum bands are currently used it will be necessary to extend the framework set out in the pricing chapter above (section 7.5) to take account of further interdependence between the uses of bands for different purposes.
[10] The results would not be affected if producer surplus were included as well. For further discussion, see section 12.2 below.

Table 11-2. Value to society in Assignment 1

Assignment 1	Mobile 700 MHz	Broadcasting 10 GHz	Total
A. Cost	100	120	220
B. User benefit (gross)	400	200	600
C. External benefit	200	500	700
Value to society (– A + B + C)	500	580	1080

Table 11-3. Value to society in Assignment 2

Assignment 2	Mobile 2.3 GHz	Broadcasting 700 MHz	Total
A. Cost	200	80	280
B. User benefit (gross)	400	200	600
C. External benefit	200	500	700
Value to society (– A + B + C)	400	620	1020

Assignment 2 (i.e. 1020). The reason for this is that Assignment 1 results in a reduction in the aggregate cost of the broadcasting and mobile services of 60 (i.e. 280 – 220) compared to Assignment 2, whilst the aggregate user and external benefits are the same in both assignments.

This illustrates the important principle that alternative spectrum assignments should be assessed in terms of the "opportunity cost" of the band in question measured by the increase in costs which would be incurred if a second-best band were used instead (see Chapter 7). As can be seen in Table 11–4, the opportunity cost of the 700 MHz band (200 – 100) is higher for mobile than for broadcasting (120 – 80) by an amount that reflects the difference in cost savings. Opportunity costs are decisive because there is no change in user or external benefit associated with switching frequencies for either broadcasting or mobile services.

If there were such differences, they would have to be inserted in rows B and C in Table 11-4, and the efficient allocation would be determined by comparing the new "Value to society" entries in that row.

11.7.2 No suitable alternative spectrum for broadcasting service

Next, suppose that there is no suitable alternative spectrum to the 700 MHz band for the broadcasting service, so that it would not be provided under Assignment 1. In this scenario, assigning the 700 MHz band to mobile eliminates the broadcasting option entirely, reducing its costs and benefits to zero.

Table 11-4. Incremental value of the 700 MHz band to each service

Incremental value of 700 MHz spectrum	Mobile 2.3 GHz	Broadcasting 10 GHz
A. Opportunity cost	100	40
B. User benefit	0	0
C. External benefit	0	0
Value to society (A + B + C)	100	40

Table 11-5. Incremental value of 700 MHz band to each service

Incremental value of 700 MHz spectrum	Mobile 700 MHz	Broadcasting 700 MHz
A. Cost	100	−80
B. User benefit (gross)	0	200
C. External benefit	0	500
Value to society (A + B + C)	100	620

Against this counterfactual, the incremental costs and benefit of assigning the 700 MHz band to broadcasting are as shown in the right-hand column of Table 11-5. Assigning the 700 MHz spectrum to mobile only generates a cost saving. Assigning it to broadcasting imposes a cost penalty, but yields large benefits.

11.7.3 The broadcasting service is not financially viable using 10 GHz spectrum

Returning to the initial example, suppose that broadcasting service revenues are not sufficient to cover the higher service costs incurred using 10 GHz spectrum. If this is not addressed in some way, broadcasting services using 10 GHz spectrum would not be financially viable and hence would not be provided.

This market failure could be remedied by providing an appropriate subsidy to increase broadcasting revenues so as to ensure that use of the 10 GHz spectrum is financially viable. This would allow spectrum efficiency to be maximized, with the 700 MHz band assigned to mobile, and the 10 GHz band to broadcasting.

If such a subsidy were not available, then the second-best outcome would be to assign the 700 MHz band to broadcasting. In effect, spectrum efficiency is distorted in order to subsidize broadcasting services. If the benefits from broadcasting on 700 MHz can be partially replicated by other means, then there is a trade-off which must be assessed between the incremental cost to society to using the 700 MHz band "inefficiently" and the loss of benefit from the full 700 MHz broadcasting service.

11.7.4 Will an auction of the 700 MHz band result in the efficient outcome?

Table 11-4 shows that, in our initial scenario, mobile operators stand to gain more than broadcasters from getting access to the 700 MHz band, as against the assumed alternative for each. Their willingness to pay would exceed that of broadcasters, and so they would win in a second-price auction.

The auction might be inefficient if there were an increase in the external benefit provided by broadcasters when the service was offered at 700 MHz as against the relevant alternative of 10 GHz, and there was no equivalent difference in external benefits provided by mobile operators. By definition, an operator's willingness to pay for spectrum does not reflect levels or changes in the external benefits it confers. If the external benefit conferred by broadcasters changed by virtue of the service being delivered using 700 MHz rather than 10 GHz spectrum, it might on our assumptions become efficient for them to have the 700 MHz spectrum, even if they would be outbid for it in an auction.

11.7.5 Conclusion

This analysis shows that there is a complex "translation" process involved in deriving inferences concerning spectrum assignment from the costs and valuation of spectrum-using services. Essentially this is because, in the allocation of an input, both cost and demand factors are involved. The relative importance of the two aspects depends upon (a) the availability of alternative ways of delivering the spectrum-using service, (b) the degree to which alternative ways of delivering the service produce the same benefits for customers and external benefits, and (c) the ability of policy makers or regulators, when a market failure arises in services markets, to intervene in those markets (which is the best approach) rather than in input markets (which is second-best).

The analysis shows that, if competing services have alternative means of delivery producing the same outputs, and the same external benefits, and if at least one of those means is commercially viable, then a competitive auction will generate an efficient outcome.

The focus on the cost side of the equation is illustrated by Ofcom's cost–benefit analysis of changing the use of the 700 MHz band in the UK to mobile services. This effectively held constant the output of broadcasting and other services competing for the band with mobile communications. It then showed that the costs of repackaging these services into spectrum in the 470–694 MHz band or elsewhere was less than the sum of reduction in costs of expanding mobile services and the benefits in better building penetration and rural coverage by mobiles associated with use of the 700 MHz band as compared with the relevant alternatives [14]. Thus with the change in outputs confined to fairly limited changes in quality of service, most of the calculation revolved around costs. However, it will not always be possible to confine the allocation problem in this way.

11.8 The struggle for the UHF band: the options

Terrestrial broadcasting requires access to radio spectrum – and with the current configuration of towers and viewing antennas that spectrum must be in the UHF band (roughly in the region of 400–800 MHz). Broadcasters would like more spectrum to enable them to deliver more content at higher definition. However, there is much competing demand for this spectrum from mobile-telephony providers. This is because the signal from a mobile network has a longer range the lower the frequency used. Roughly, each doubling of frequency results in a halving of range. Since the coverage area of a cell is related to the square of the radius, a halving in range requires four times as many cells to achieve the same coverage. Compared to some recently released cellular spectrum at, say, 2.6 GHz, the TV spectrum at 600 MHz would only need one-sixteenth as many cells. This is not the whole story because of issues such as antenna size, man-made noise, and in some cases the need for more cells for capacity reasons. Nevertheless, there are clearly strong advantages for mobile operators in using lower frequencies in the UHF range. This is shown by the results of spectrum auctions: a megahertz of spectrum at 700 or 800 MHz generates revenues many times greater than a megahertz at 1800 MHz or 2.6 GHz.

Thus there are many operators who would like access to the UHF band. These include:

- Mobile operators. They prefer a portfolio of spectrum with some low-frequency (e.g. UHF) spectrum to enable cost-effective coverage in rural areas and some high-frequency spectrum to provide high capacity in urban areas. At present, they have more high-frequency than low-frequency spectrum and some are looking to increase their low-frequency holdings in order to expand their coverage. This also assists some governments in their targets to provide broadband coverage to the vast majority of the population (see Chapter 13 below). Their need for spectrum has been exacerbated by the dramatic growth in mobile data usage, discussed in Section 11.5.
- Public protection and disaster relief (PPDR) – also known as emergency services. Many police, fire, and ambulance users already have systems in the UHF bands either at 400 or 800 MHz. These systems are typically voice-based and the users would like to expand them to deliver content at high data rates. For this they will need additional spectrum, and if this can be offered in bands close to their current systems they will be able to continue to use the same transmitter sites. In the US, one of the reasons given to the public for digital switch-over of TV transmission was to free up spectrum for PPDR to enable better response in emergency situations such as the 9/11 terrorist attacks. The projected US auction of 600 MHz spectrum, the so-called "incentive auction" discussed in Section 4.9, is intended to gain revenues sufficient to fund the construction of a broadband network for the emergency services. It is also possible for the emergency services to use capacity on a commercial network, provided that an arrangement is in place to give these users pre-emptive access to capacity in the event of a major emergency. This solution is being investigated in some jurisdictions.

- Machine-to-machine (M2M) systems designed to link sensors back to control systems. Typically the sensors need to be low-cost and to run for many years off a single battery. To do this they need to be able to transmit with low power levels, which would normally result in a short range, but the superior propagation characteristics of the UHF band mitigate this and make deployment practical. Technologies such as Weightless are predicated on gaining access to the UHF TV spectrum, albeit on a secondary basis as dedicated spectrum is not available.

Over recent years the trend has been toward a steady erosion of the TV band. In Europe the band previously extended to 868 MHz. During digital switch-over, the "800 MHz" band from 792 to 868 MHz was cleared and auctioned for mobile use, and it is now part of the 4G band around the world.

In the US, which has a different broadcast configuration, the 700 MHz part of the band was cleared during digital switch-over and auctioned off to cellular operators. Many are now deploying 4G solutions in this band. This has led to widespread availability of equipment, particularly handsets, prompting other countries to consider whether they should also clear 700 MHz. This issue was raised in the 2012 World Radio Conference (WRC) [15], where there was agreement that cellular use in the 700 MHz band should be enabled worldwide. While this does not place an obligation on regulators to clear the band of broadcasting, many regulators took it as a sign that they should at least consider the future of this part of the band, potentially restricting TV broadcasting to 470–700 MHz. A number of regulators are now studying the implications of this and publishing consultations on the future of the 700 MHz band.

The broadcasters, understandably, see this as a potential trend whereby they are increasingly compressed into ever-smaller parts of the spectrum. They rightly point out that at some point there will be insufficient spectrum for a commercially viable service and they might as well switch off terrestrial broadcasting altogether. In some countries where there is already very high penetration of cable or satellite viewing (e.g. Germany) this might be politically acceptable and commercially sensible. However, in others (e.g. the UK, where over half of households still rely on DTT) there would be extremely strong opposition to such a plan. But given the international co-ordination necessary while some broadcasting remains in the band, the ability for each country to make an independent choice is limited, resulting in much complexity.

In the US, they are trying to make use of market forces to find a way through these complexities. The US, like Germany, has relatively little reliance on DTT, with only around 10% of households using it as their primary service. They are planning a novel "incentive auction" in 2015. This has been described in detail in section 4.9 above, and uses an auction mechanism to determine how much spectrum is retained for TV broadcasting and how much is switched to mobile operators. The process will be closely watched by other regulators, both to see if the approach works and as an indication of whether the demand from cellular operators for spectrum below 700 MHz is high.

11.9 Possible outcomes

We can imagine a number of different outcomes, such as:

- the status quo, where the remainder of the band (around 470–790 MHz) remains for broadcast use for the foreseeable future (noting that in some countries the 692–790 MHz band has already been promised for mobile use);
- the converse, where broadcasting vacates the band and the entire band is used for mobile broadband or some other new service such as M2M;
- an intermediate position where some additional spectrum is released from broadcasting, leaving the band divided between broadcast and mobile use; and
- a hybrid approach where a single network delivers mobile and broadcast content in a flexible and dynamically changing manner.

It is possible that different countries will aim for different outcomes depending on their use of terrestrial broadcasting and their perceptions of the values of broadcasting and mobile. Equally, global economies of scale tend to bias much of the world toward similar use of spectrum and the complexities of frequency planning across borders in these bands will tend to mean that neighboring countries will adopt similar usage of spectrum.

Some of the key factors that will impact on the outcome selected include

- whether there is continued growth in mobile data demand and how operators choose to meet this,
- whether viewing patterns change in the future, and
- governmental priorities in areas such as mobile broadband provision.

The impact of these factors is not always obvious. For example, a continued rapid growth in demand in mobile data would require a different approach to capacity enhancement than adding additional spectrum as this can only increase capacity by some tens of percent. Instead, a massive number of small cells might need to be deployed. Small cells work better at higher frequencies since they do not require the range advantages conferred by the UHF band. Hence, operators might be more interested in spectrum at, say 3.5 GHz than at 700 MHz as a result. Equally, an increase in the demand for high-definition and ultra-high-definition TV viewing would result in there being insufficient capacity on the UHF multiplexes to deliver the content and viewers might instead migrate to platforms such as satellite, reducing the demand on UHF. Perhaps the most complex case would be one which combined a steady growth in mobile broadband that could be met with more spectrum and a moderate decline in broadcast viewing which fell short of justifying a switch-off of terrestrial broadcast transmission.

Many have assumed that since the 800 MHz band has moved from broadcast to mobile, and the indications are clear that this move will be repeated at 700 MHz, then it is only a matter of time before the 600 MHz and then the 500 MHz band follow along the same path. However, recent work for the EC suggests that the interest from mobile operators in frequencies below the 700 MHz band may be relatively low.[11] Operators

---

[11] See the study led by Plum Consulting at [16].

currently appear to believe that they may have access to sufficient UHF spectrum at this point and that the marginal gains of additional lower frequencies will be small. Instead, they are using their regulatory teams to lobby for increased availability of higher-frequency spectrum. Broadcasters appear determined to continue their transmission in the remainder of the band, and at present the public demand for linear TV broadcasting remains strong. Governmental interest in forcing a change in viewing habits is low. Hence, the status quo appears the most likely outcome, at least for the next decade.

11.10 Implications for spectrum management

The issue described here is probably the area of sharpest conflict in spectrum management at present. It encapsulates all the major points of controversy, including how fast demand for mobile data is growing, the long-term prospects for terrestrial broadcasting, the value of economic externalities from broadband, the value of social externalities from broadcasting, technical progress in both broadcast and mobile radio technology, and even the possibilities that the services will converge onto a single platform in the future.

It is impossible to resolve these conflicts in a way which will generalize from one country to another. However, we can seek to derive some conclusions about the direction of travel of spectrum management from the example.

First, countries differ in the tools which they use to resolve this issue. Most are falling back on command-and-control methods, preceded by cost–benefit analysis or regulatory impact assessment. Only in rare cases is recourse had to market methods such as trading or a service-neutral auction. The best example of this approach is provided by the "incentive auction" proposed in the United States for the 600 MHz band.

Explanations for command-and-control methods overriding, to some extent, economic tools such as trading or a full auction are several in number:

- the need for haste to avoid the spectrum crunch,
- the need to harmonize the spectrum transition across countries,
- the complexity of market allocation and assignment processes,
- the multiplicity of external effects involved and the problem of incorporating them in the auction mechanism,
- the politicized nature of the decision and the desire to be seen to be taking an active role,
- the sharpness of the conflict between the broadcasting and mobile sides, and
- governments' desire to maximize auction revenues.

It is difficult at the moment to detect whether, in future, other complex and controversial spectrum decisions will default to command and control.

References

[1] See www.itu.int/ITU-R/terrestrial/broadcast/plans/ge06.
[2] See http://en.wikipedia.org/wiki/Multimedia_Broadcast_Multicast_Service.

[3] See http://europa.eu/rapid/press-release_IP-14-14_en.htm.
[4] See TG6(13)026 by IRT, available from the ECC website.
[5] See Broadband Commission, "The State of Broadband 2014: Broadband for All," ITU/UNESCO (2014).
[6] J. Stewart, "Mobile Broadband Coverage: Balancing Costs and Obligations," at www.analysysmason.com/About-Us/News/Newsletter/Mobile-broadband-coverage-balancing-costs-and-obligations.
[7] See www.cisco.com/c/en/us/solutions/collateral/service-provider/visual-networking-index-vni/white_paper_c11-520862.html.
[8] See www.whitehouse.gov/the-press-office/presidential-memorandum-unleashing-wireless-broadband-revolution.
[9] J. Genachowski, "Winning the Global Bandwidth Race: Opportunities and Challenges for Mobile Broadband," University of Pennsylvania (October 4, 2012), at www.fcc.gov/document/chairman-genachowski-winning-global-bandwidth-race.
[10] P. Diamond and J. Mirrless, "Optimal Taxation and Public Production" (1971) 61 *American Economic Review* 8 and 261.
[11] See http://stakeholders.ofcom.org.uk/consultations/ddr.
[12] Ofcom, "Digital Dividend Review" (December 2007), Annex 7, 19, at http://stakeholders.ofcom.org.uk/binaries/consultations/ddr/annexes/ddr_annexed.pdf.
[13] UK Government, "Spectrum Strategy" (2014).
[14] Ofcom, "Consultation on Future Use of the 700 MHz Band: Cost–Benefit Analysis of Changing Its Use to Mobile Services" (May 2014).
[15] See ITU website, e.g. www.itu.int/ITU-R/index.asp?category=conferences&rlink=wrc-12&lang=en.
[16] See http://ec.europa.eu/digital-agenda/en/news/challenges-and-opportunities-broadcast-broadband-convergence-and-its-impact-spectrum-and.

12 Public-sector spectrum use

12.1   Introduction

In most countries a great deal of spectrum – roughly half – is used within the public sector. This applies both to all bands taken together and to the highest-value spectrum in the 300 MHz to 3 GHz range. The uses include applications for aviation and maritime transport, communications, radar, and science. Most of the spectrum use is defense-related.

The pattern of use across countries is not very sensitive to geographical location and income per head, so a similar division can be expected everywhere. It is also common for defense-related spectrum to account for more than half of public-sector spectrum use. Figure 12-1 shows the breakdown between commercial and public use of spectrum in a typical European country.[1]

Historically, defense and, to a lesser extent, other public-sector uses of spectrum had priority over civilian uses. Often public-sector users were not subject to license requirements or to license fees in the same way as commercial users. In periods of spectrum plenty (i.e. before the recent growth of spectrum demand for communications services), the public sector was normally given generous assignments.

In the UK a study monitoring spectrum use found low levels of use of some public spectrum, including in the 1.0–1.8 GHz range, as contrasted with high levels of use in mobile bands – for example 900, 1800, 2300 MHz.

This says nothing about the value of such use. However, the observation both of low usage levels and of the lack of incentive for efficient use of public-sector spectrum which is described below does at least raise questions for further investigation.

This chapter draws heavily upon the broader analysis we have undertaken in previous chapters, notably of spectrum scarcity and its consequences (Chapter 3), of spectrum pricing and valuation (Chapter 7), and of spectrum sharing (Chapter 8). First we discuss how special attributes of the public sector affect spectrum use and spectrum management. Then we review a set of possible means to improve efficiency of spectrum use there.

---

[1] Note that the dividing line between commercial and public uses of spectrum is not straightforward. For example, air traffic control – in most countries an organization in the public sector – supplies services to commercial aviation. Equally, communications for the emergency services in the public sector (police, fire, ambulance) may be provided by commercial networks.

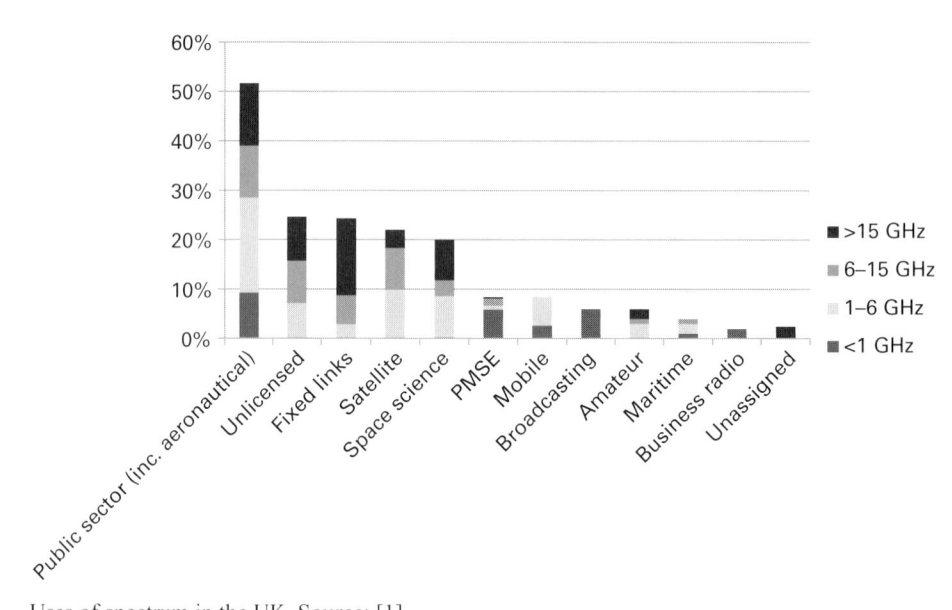

Figure 12-1. Uses of spectrum in the UK. Source: [1].

12.2 Differences between commercial and public-sector use

One of the major themes of this book has been the pressure placed upon the traditional "command-and-control" regime of spectrum management by rapidly increasing demand for spectrum for mobile communications. We have suggested that the administrative balancing act which worked satisfactorily in the period of spectrum abundance falters as the consequences of allocation mistakes increase.

This applies equally to public and to commercial spectrum. When the public sector needs spectrum to take advantage of a new technology, it needs it at once. But because of generous historical treatment by spectrum managers of public-sector bodies, the natural direction of travel for spectrum is likely to be from the public to the commercial sector. And when the public sector holds on to spectrum unnecessarily, the consequences for citizens and for the economy can be considerable.

It was, and still is, quite widely assumed that a command-and-control spectrum management regime "fits" the public sector quite well. However, this is questionable. Public-sector bodies require many inputs to produce the services they do, including land and other natural resources, labour, energy, capital equipment, and materials. With notable exceptions, such as the military draft of recruits which still exists in some countries, these inputs are not administratively allocated to public production in a modern market economy. Instead the public sector acquires them, bidding against commercial firms.

It follows from this that the same economic and market tools, described in Chapter 3 above, can be deployed with public-sector spectrum as are used with commercial spectrum. But there are two major differences. First, while private- and public-sector bodies are likely strenuously to resist having either to return spectrum, or to start paying for it, the nature of the service provided and the relative ease and informality with which certain public-sector bodies, such as defense ministries, acquired the spectrum have given them an unusually strong sense of entitlement to what they now hold. A direct assault on this deeply held feeling can often misfire. Accordingly it may be better to approach the issue in stages in a more strategic way: first by disseminating information about the value in commercial hands of the spectrum bands in question; then by persuading users that a reform of spectrum management, if implemented carefully, would not threaten public users' central interests; and then by pointing out that such a reform would increase the chance of the relevant public-sector user itself gaining additional spectrum in the bands where it needs it. It is also helpful to seek to enlist the support of the country's finance ministry, which should have an interest in applying underused public-sector spectrum to promote economic growth in the country as a whole.

Second, the regime of incentives within which the public sector works is different from that in which the commercial sector works – the difference also varying within the public sector. Commercial firms almost invariably produce a marketed output. If their costs exceed their revenues, they are under pressure to cut their costs. In other words, they are under a hard budget constraint: they cannot persistently lose money. Thus it is reasonably foreseeable that, providing a commercial firm is paying a nontrivial price for access to spectrum (i.e. not a modest administrative charge), it will try to economize on that input.

A firm in public ownership which sells a marketed output (for example, a public water or energy firm) is also likely to have financial targets, although these may not be as hard as those for a commercial firm. Accordingly, incentives for it to economize on spectrum may still be present. However, a user of public spectrum like the ministry of defense does not produce a marketed output but one which is provided free of charge for all inhabitants. It will be funded by a grant from the ministry of finance, the size of which is intended to cover its costs.

If this body uses spectrum, it will probably negotiate with the ministry of finance over its budget. It is possible to think of two different contexts within which this negotiation might take place.

A. In country A, the spectrum user submits a list of inputs for the forthcoming period, and of their prices. These are scrutinized by the ministry of finance. A budget emerges, which may be approved by parliament. It lists which inputs can be purchased. The spectrum-using department then purchases and uses the specified inputs to produce its outputs.
B. In country B, on the other hand, a different regime operates. The ministry of finance agrees output targets for the department in overall terms for a period of, say, three to five years. The two negotiate an overall budget to achieve this output. It is then up to the department concerned to spend the budget in the best possible way to achieve the outputs specified.

In country A, the impact of an increase in the price of spectrum may turn out to have little effect on the efficiency of spectrum use. It may simply trigger an increase in the budget allocated to the spectrum-using department by the ministry of finance. Moreover, if the department saves spectrum, it may not be able to reassign the value of that saving to acquire another input. In this case, there is no incentive to respond to price changes. In country B, by contrast, if the department saves money on spectrum use it can redirect it to other purchases. In this case, there is an incentive to respond to price changes.

It should be clear that the problem of encouraging efficient spectrum use in the non-marketed public sector is not confined to spectrum alone. It is common to the use of all inputs and production of all outputs by all public-sector departments producing non-marketed outputs. The regime described as relating to country A corresponds in general terms to traditional budgeting practices in the public sector. The regime described as relating to country B corresponds in general terms to what is recommended by some adherents to what is called the "new public management" [2], which seeks to improve efficiency in management across the whole public sector. As shown below, this regime seems well or better equipped to achieve spectrum efficiency by creating appropriate positive incentives for spectrum users than the rival approach.

If, however, the approach used in country B is impracticable, some of the methods discussed below, including, notably, refarming and the use of audit, are still available.

12.3 A program of reform of public spectrum use

This section discusses a range of possible tools for improving public-sector spectrum use. Some of them are the same as, or developments of, traditional methods. Others are the same as market- or price-based reforms of more general application, sometimes adapted to meet the special circumstances of the public spectrum users. The "ladder" of public-sector reform shown in Figure 12-2 describes from rung to rung a succession of traditional and reforming actions which the spectrum regulator or government may want to consider, not necessarily in the order shown, as a program to achieve greater efficiency in the use of public-sector spectrum. It is based heavily on UK experience.

Step 1: subsidized refarming Refarming is the process of changing the use of a band.[2] In the present context, it means taking spectrum from a public-sector user and assigning it to commercial use. If the band assigned to the public-sector user were no longer used at all, technically it would be relatively straightforward. But complications would arise if it were still in use (as it would have to be cleared) or if it were intended for future use. In either of these cases some form of compensation might be appropriate.

Procedures for refarming spectrum are particularly well developed in France. When a band has been identified as a candidate for refarming, a careful evaluation of the costs is made, involving regulators and the parties leaving and moving into the band. With ministerial approval, the spectrum regulator, Agence nationale des fréquences, approves payments out of a redeployment fund, which receives public funds and also contributions from the commercial bodies moving into the frequency.

---

[2] This change can be accomplished by secondary trading without any regulatory participation. However, the discussion of subsidized refarming here does not consider this mode.

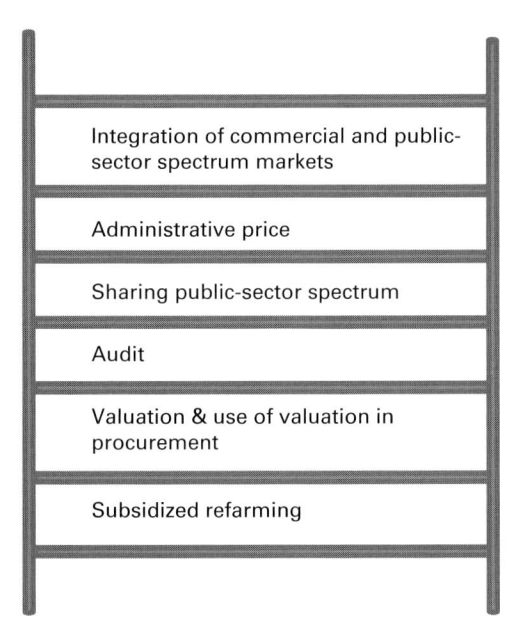

Figure 12-2. The ladder of public-sector spectrum reform.

The French parliament has now set up a special-purpose account, which has been used since 2009 to disburse funds to organizations giving up frequencies, in many cases recouping them from subsequent auction revenues. The majority of spectrum that has been freed up for broadband mobile so far has come from the ministry of defense. It vacated the 1800 MHz band, a big part of the 2.1 GHz band, the E-GSM spectrum, the 2.6 GHz spectrum, and half of the 800 MHz band. In 2011, the Ministry received $1.2 billion – the full amount of auction receipts of the 2.6 GHz band, and in 2012 it received an additional $1.8 billion for releasing part of the 800 MHz band [3].

The French government also intends to use the fund to smooth the release of the 700 MHz band. In this case, however, there is no release of spectrum from the defense ministry, as all the spectrum comes from the broadcasting band. But both broadcast spectrum and government spectrum are managed in similar centralized ways.

The US also developed, and has applied on one occasion, an interesting market-guided procedure. In 2004, the US Congress passed the Commercial Spectrum Enhancement Act, which created a Spectrum Relocation Fund, financed out of auction proceeds, to cover the costs incurred by US federal government entities which relocate to new frequency assignments or alternative technologies. The basic idea is that if expected auction revenues exceed expected transfer costs, the transfer should take place. It is possible to use the expected transfer cost as the "reserve price" in the auction. If that price is not attained, the sale does not take place.

The first auction of this kind to occur was of the 1710–1755 MHz band, which was used by 12 federal agencies, including the Department of Defense. The sale was accomplished, and a portion of the auction proceeds was used to cover spectrum relocation costs. It subsequently transpired that these costs exceeded estimates by about 50%. However, these larger costs were still exceeded by auction revenues, by more than four times [4]. A similar process has been contemplated in the US with the 1755–1850 MHz band.

Step 2: use of spectrum valuations in public-sector procurement The hypothesis is that public-sector spectrum may be used inefficiently because it is free and users do not appreciate its value. Estimating the aggregate value of spectrum employed, for example, in defense uses, can show government departments and lawmakers the billions of dollars of assets which are locked up in public-sector spectrum uses. This can be part of the educational process of encouraging economy in spectrum use.

Chapter 7 has shown that there are several methods of estimating the value of spectrum and of setting administrative prices for its use; these can be applied to any frequencies. It is possible initially to use such prices not in their normal role of determining how much money has to be paid annually by the spectrum user, but as an aid to making other decisions. Used in this fashion, they are "shadow prices" – i.e. prices which are used in making an investment decision but not actually charged.

For example, if the government is deciding between two weapons systems, using different spectrum bands, the appraisal can be based on the lifetime costs of the two systems, including the valuation of the spectrum each would employ. The expected result is a better choice of weapons system and a better understanding of spectrum's value.

If this is to be accomplished, it will probably require the endorsement of the finance ministry, which in many countries lays down rules which govern the process of public-sector procurement.

Step 3: audit Spectrum audits are not so much an economic or market tool in spectrum management as a helpful first step or even a prerequisite for the use of such tools – since they enhance knowledge of the scale of use of the various bands, on the part both of the government and regulator and of spectrum users themselves.

The UK government commissioned an audit of spectrum use published in 2006, which identified certain underutilized bands in the public sector [5]. This audit also set out some general principles for management of public-sector spectrum, many of them discussed in this chapter. It also audited twenty specific bands managed by the Civil Aviation Authority and the Ministry of Defence. Examining current and potential future use, the audit took the view that in eight bands there was scope for making alternative use of the band in the short term. In three bands, there was little scope for alternative use in the next five years. In thirteen bands, potential for some redeployment was identified, either after further examination or over a longer time period.

Accepting the recommendations, the UK government noted that the latter two categories accounted for 65% of public-sector spectrum. A series of actions was agreed to examine bands capable of alternative use [6]. Subsequent developments are discussed in section 12.4 below.

In 2013 in the European Union, as part of the Radio Spectrum Policy Programme, the European Commission published a decision to implement a spectrum inventory within the EU [7]. Such audits cannot be comprehensive, as some spectrum uses have to be kept secret for reasons of national security; however, the two examples show that useful information about public-sector spectrum use can safely be disclosed.

Step 4: sharing public-sector spectrum Much attention has recently been paid to the sharing of public-sector spectrum with the commercial sector, especially for the purposes of providing mobile broadband. In 2012, the US President's Council on Science and Technology published a report which advocated freeing up 1,000 MHz of government spectrum for shared use. Today's apparent shortage of spectrum, it is claimed, is in fact an illusion created by the way the resource is managed. If the US widens its options for managing federal spectrum, spectrum availability will be transformed from scarcity to abundance. The norm for spectrum use should be sharing, not exclusivity.

As explained in section 3.4 and Chapters 8 and 9 above, sharing can take many forms. The simplest ones involve sharing by geography or by time. Examples are fixed links sharing with satellite uplink transmissions. More complex forms of sharing involve giving the public sector some form of pre-emption right over a second, commercial user. Then there is dynamic spectrum access of the kind described above in Chapter 9.

In November 2013 it was announced that the US Department of Defense (DoD) has agreed to vacate the 1,755–1,780 MHz band to allow it to be paired with the 2,155–2,180 MHz band in an auction. It has previously been estimated that the paired block would raise nearly $12 billion, while the worth of the 2,155–2,180 MHz band by itself has been estimated at just $3 billion. The military currently uses the 1,755–1,780 MHz band for pilot training and the operation of drones. These technologies will move to the 2,025–2,110 MHz band, which will be shared with broadcasters in accordance with an agreement it has made with the National Association of Broadcasters (NAB).

However, others cast doubt on the practicability of public/commercial spectrum sharing, which requires not only the solution of technical problems, but also appropriate incentives for both parties and control over transaction costs. As with other sharing proposals, a sensible approach in other countries may be to monitor the success of such ventures in early-adopting countries and be prepared to follow if this turns out to be a successful and economic strategy.

Step 5: charging spectrum prices The step up from setting shadow prices for spectrum (see step 2 above) to charging actual administrative prices is a major one, which is strongly resisted by both commercial and public-sector spectrum users.

As noted above, if the ministry of finance sets a fee which a government department must pay for spectrum, and simultaneously and automatically increases that department's financial allocation by exactly the same amount, the system will not work, as there is no incentive to economize.

Accordingly, spectrum pricing only works within a system for managing public expenditure in which the ministry of finance makes an overall allocation of funding to the relevant department against specified output targets, leaving the department to decide how to meet those targets. In such circumstances, a department will choose to hand back unwanted spectrum and use the money saved more effectively elsewhere. In these conditions, spectrum pricing can promote the return of unutilized spectrum, as the UK case studies below indicate.

We noted in section 7.5 above that opposition to administrative prices is to be expected. There are several ways of seeking to defuse opposition to public-sector charges. As described above, a period of shadow operation can precede the introduction of charges, the level of charges can be gradually increased from a low base, and the ministry of finance can be persuaded to issue a policy statement on how it will take account of such charges in public spending allocations. It can also be linked with increased power for the public-sector body to realize some value from its spectrum holdings, as described below.

Step 6: integrating commercial and public-sector spectrum markets As noted above, in market economies, most inputs (labour, land and buildings, utility services, capital equipment) are bought by public-sector organizations in broadly the same way as by commercial organizations – in an integrated marketplace. Spectrum has been a historical exception in being largely assigned by administrative processes. But when a market for commercial spectrum is created, it is natural to ask whether it should include public-sector spectrum, with the two types of organization vying with each other for the spectrum which they need, as they now compete for labour and materials.

In such a world, if additional public-sector spectrum were needed, the relevant organization would have to seek funds from the ministry of finance to acquire the spectrum – at auction, for example. Equally, if a public-sector organization had surplus spectrum, it could either give it back or auction it on the market. If the seller could keep some of the proceeds, this would encourage it to sell.

For the reasons given above – the historically generous treatment in spectrum assignments of the public sector and the fact that a mainly commercial activity, mobile communications, is the major source of current additional spectrum demand – the direction of flow of spectrum in the future is likely to be from public to commercial uses. Here a key issue is the willingness of the ministry of finance to allow a public-sector body to keep a proportion of revenues from the sale or lease of spectrum sufficient for it to be willing to forgo the "cushion" of keeping spare spectrum. This financial incentive can, of course, be supplemented by making release of spectrum a high-priority policy goal, as is described in the next section for the case of the UK, where a trade of spectrum across the public–commercial boundary is planned.

Achieving integration with respect to auction bids or purchases of spectrum by public-sector bodies presents greater problems. As described above, it requires an allocation of funds by the ministry of finance to make the purchase. It is likely that this will only be granted after a "value-for-money" test, which public-sector bodies have little experience of with respect to spectrum. In the UK, as elsewhere, there has been long agitation, supported by equipment manufacturers, for additional spectrum to be made available by a traditional 'command-and-control' process to supply broadband for the emergency services. This may have got in the way of the construction of a business case, which would have had to consider the option of using a commercial supplier. Thus the integration only works when the public sector accepts it as the way in which spectrum is assigned.

12.4 An example of public-sector spectrum reform: the UK

The United Kingdom is one of the countries which have been engaged in the reform of public-sector spectrum management, and its experience has strongly influenced the discussion above.

The chronology is set out in Box 12-1.

Box 12-1. Chronology of UK public-sector spectrum management reforms

1998: beginning of administrative scarcity prices [8]. The 1998 Wireless Telegraphy Act introduced in a limited number of commercial bands administrative pricing for spectrum which took account of scarcity. Extension to public spectrum was also seen as appropriate.

2003: passage of Communications Act 2003.

2005: Ofcom publishes the Spectrum Framework Review [9], which adopted an illustrative target of managing 71.5% of UK spectrum using market methods by 2010.

2005: independent spectrum audit [6] inaugurates program of reviews of bands. Government adopts policy of extension of prices and integration of commercial and public spectrum markets.

2010: Ofcom review of administrative prices [10] confirms usefulness of spectrum pricing and confirms applicability to public-sector spectrum.

2010: extension of administrative prices to more bands. The Ministry of Defence annual bill nears £400 million.

2010: government announces plan to release 500 MHz of spectrum below 5 GHz by 2020 [11]. The plan was signed by eight ministers, and 60 MHz was released by early 2014. One hundred MHz are to be released by the Department for Transport.

2015: plans are made for 190 MHz of spectrum held by the Ministry of Defence to be auctioned on its behalf by Ofcom [12].

12.5 Conclusion

Improving efficiency of spectrum use in the public sector involves greater complexity than in the private sector because the public sector has more opaque incentives and its operations hinge to a large degree on its method of financing. In principle, commercial and public spectrum markets can be integrated, to the advantage of both, but in practice this is certainly difficult to accomplish in the short term and, without a broad culture change in the public sector, it may be too much to ask in the longer term as well.

There are a number of messier tools available. These include spectrum prices, where the planning of public spending supports them; audits; use of governmental pressure; and administrative refarming with some form of compensation. It seems likely that much can be done by a combination of these methods.

References

[1] Ofcom, "Spectrum Attribution Metrics" (December 2013).
[2] T. Christensen and P. Laegreid, eds., *The Ashgate Research Companion to New Public Management*, Farnham: Ashgate, 2011.
[3] See "Mission 'Défense' et Compte d'affectation spéciale 'Gestion et valorisation des ressources tirées de l'utilisation du spectre hertzien'," Sénat, République de France, at www.senat.fr/commission/fin/pjlf2014/np/np08/np081.html.
[4] See www.gao.gov/assets/660/654794.pdf.
[5] M. Cave, "Independent Audit of Spectrum Holdings" (2005), at www.spectrumaudit.org.uk.
[6] Independent Audit of Spectrum Holdings: Government Response and Action Plan (April 2006), at www.spectrumaudit.org.uk/pdf/governmentresponse.pdf.
[7] See http://eurlex.europa.eu/LexUriServ/LexUriServ.do?uri=OJ:L:2013:113:0018:0021:EN:PDF.
[8] Radio Communications Agency, "Spectrum Pricing: Third Stage Update and Consultation" (December 2000).
[9] Ofcom, "Spectrum Framework Review" (June 2005), at http://stakeholders.ofcom.org.uk/binaries/consultations/sfr/statement/sfr_statement.
[10] Ofcom, "SRSP: The Revised Framework for Spectrum Pricing: Our Policy and Practice of Setting AIP Spectrum Fees," statement (December 2010).
[11] DCMS/Shareholder Executive, "Enabling UK Growth: Releasing Public Spectrum: Making 500 MHz of Spectrum Available by 2020," Public Sector Spectrum Release Programme update (March 2014).
[12] See http://stakeholders.ofcom.org.uk/binaries/consultations/2.3–3.4-ghz/summary/2.3–3.4-ghz.pdf.

13 Spectrum and the wider economy

13.1 Introduction

Spectrum management used to be an esoteric specialism practiced mainly by engineers outside the knowledge and understanding of policy makers, businesses, and consumers. That all changed with the rapid diffusion of mobile communications in the 1990s, and the consequential rocketing in the values of spectrum, vividly illustrated in the auctions of 3G spectrum around the year 2000. Since that time, spectrum management has risen in importance and, as part of this process attention has been directed to the growing role of spectrum-using services in the economy and in society.

In the case of some natural resources, calculations of the level of economic activity they add to the economy are justified by the possible exhaustion of supply. In the case of spectrum, this is not an imminent or even a distant prospect. However, estimates of the growing importance of spectrum-using services do underline the importance of allocating what we have efficiently. Thus if misallocation and hoarding reduce the effective stock of spectrum by half, a quick and inexpensive way of increasing output and welfare in the economy is to improve the spectrum management regime.

This chapter brings together some estimates of the links between the use of spectrum and the wider economy. The focus is not on spectrum itself (as this is an input) but on the spectrum-using services which generate final output. We first consider how key services which rely on spectrum contribute to economic welfare, examine the weight of those services in gross domestic product (GDP), and consider evidence on how investment in information and communications technologies (ICT) has contributed to increasing productivity throughout the economy.

However, the impact of certain spectrum-using technologies, notably mobile voice and data communications, has the potential to pervade the economy, not only affecting firms and households directly purchasing spectrum-using services, but changing business models, altering competitive structures, and promoting innovation. Attempts have been made to estimate these external effects by relating changes in GDP to the diffusion of fixed and mobile communications services. We discuss the implications of these results.

13.2 Spectrum, spectrum-using services, and their impact on welfare

A standard way of assessing the contribution of spectrum-using services and thus of spectrum to the economy is to use the technique of economic analysis illustrated in

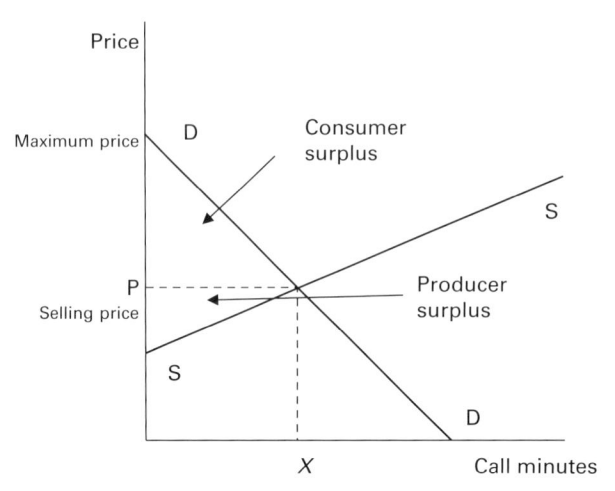

Figure 13-1. The consumer and producer surplus generated by a spectrum-using activity.

Figure 13-1, which presents an application to mobile voice telephony. The curve marked DD is the demand curve for calls in the country in question. Its downward slope reflects the fact that as price falls, the demand for calls increases. It is important for what follows that the demand curve intersects the vertical axis at what can be called the "maximum price." This is the price at which demand goes to zero or is choked off – also called the "choke price."

The supply curve (SS) represents the number of call minutes which the mobile operators would jointly produce at different prices. Its upward curve can reflect the fact that if the price per minute rises, operators will increase supply, for example building out their networks into more sparsely populated and costly areas.

The equilibrium of this market occurs at a "selling price" of P and a quantity of call minutes of X.

Now consider the triangle marked "consumer surplus." This area arises because the caller values the first minute of calls purchased at a maximum price, which is considerably in excess of the selling price. That consumer is therefore getting a benefit or surplus in excess of the price paid. That surplus diminishes down the demand curve as the quantity purchased rises to X, reaching zero at X itself. Total consumer surplus can therefore be estimated as the area shown, which on these facts, because the demand curve is linear, is a triangle.[1] The procedure requires that there be a maximum price – if the demand curve never met the vertical axis, the area of consumer surplus would not be defined.

Parallel to consumer surplus, there is a benefit to producers known as producer surplus, also shown in Figure 13.1. The shape of the supply curve reflects the marginal costs which mobile operators collectively incur in supplying call minutes. Its upward slope can be thought of as reflecting rising additional costs incurred in supplying additional minutes.

As a result of this shape, producers of some calls find themselves being able to sell units at a price (the selling price) which exceeds their marginal costs of production. This

---

[1] The expression for the area of a right-angled triangle is: ½ × height × base. If the demand curve were nonlinear, the area could be estimated using a different formula.

Table 13-1. Welfare effects (consumer and producer surplus) of seven spectrum-using services

	2011 value (£ billion)	Real increase 2006–2011 (%)
Public mobile communications	30.2	16%
Wi-Fi	1.8	N/A
TV broadcasting	7.7	79%
Radio broadcasting	3.1	35%
Microwave links	3.3	−29%
Satellite links	3.6	7%
Private mobile radio	2.3	25%
TOTAL	52.0	25%

Source: [1].

is not necessarily pure profit, as some of it may be spent on costs which do not vary with output – the cost of a mobile license, for example. It is generally known as "producer surplus."

There is disagreement over whether economic welfare should include consumer surplus alone, or should be the sum of consumer and producer surplus; underlying this issue is a question whether the profits from businesses should count as contributing to welfare.[2] Some of the welfare measures reported below are of consumer surplus alone; some also include producer surplus.

It makes sense, in the case of services which depend crucially upon spectrum as an input, to ascribe the surplus (however defined) to spectrum. This applies in the case of mobile communications, which cannot operate without spectrum as an input. It makes less sense in the case of services where spectrum plays a subordinate or even a dispensable role. Accordingly, calculations of spectrum-related welfare typically concentrate on a limited number of particular activities.

As an example of calculations of this kind, in 2012 the UK government commissioned Analysys Mason, a consultancy, to conduct a study which estimated the welfare effects of seven notable spectrum-using sectors in 2006 and in 2011. The results are given in Table 13-1. The 2011 figure as a percentage of GDP is 3.4%.[3] Some 80% of the benefit comes in the form of consumer surplus, the remainder as producer surplus.

13.3 Effects of spectrum-using services on GDP and employment

An alternative measure is via the contribution to national income, output, or product made by spectrum-using sectors. Gross domestic product is found by adding together the

---
[2] This is a hotly contested issue but we do not consider it furthur here.
[3] This figure is given just as an indication of scale, as data on consumer and producer surplus use a different metric than that used for GDP, which is described in the next section.

Table 13-2. The direct and total contribution of mobile to the Australian economy in 2011–2012

	A$ (billion)	Thousands of employees	% of GDP	% of employment
Values				
Direct contribution	7.6		0.5	
Total contribution	14.1		0.9	
Employment				
Direct effect		22.3		0.2
Total effect		56.9		0.5

Source: [2].

value added (revenues minus material inputs) of all the sectors in the economy. It is also a measure of the income available for distribution to factors of production – labour, capital, and natural resources, including spectrum. It is therefore possible to gather data on value added in chosen spectrum-using sectors and aggregate them to produce an overall figure of direct contribution to GDP.

As before, this procedure makes more sense in application to activities which rely in some fundamental and unavoidable way on spectrum, rather than in cases where the role of spectrum as an input is small or contingent.

This direct measure only captures the immediate effect on gross domestic product of the chosen spectrum-using sectors. But spectrum's impact on production is felt not only in mobile communications, but also in activities such as the manufacture of smartphones and the construction of towers to support antennas. The calculation can thus be extended to include these indirect effects. The direct and the overall (direct and indirect) impacts of spectrum on job creation can also be estimated from employment data by sector.

These calculations are illustrated in a 2013 study by Deloitte and Access Economics of the impact of the mobile sector on GDP and employment in the Australian economy from 2009–2010 to 2011–2012. Both the direct and overall contributions, and direct plus indirect contributions, were computed. The results for the final year are shown in Table 13-2.

13.4 Effects of spectrum-using services on productivity

A further extension of the above-noted analysis of indirect effects takes us into the territory of how a spectrum-using service such as mobile communications might increase productivity when it is used not as a consumer service but as an input in another sector of production. This calculation is of interest as a component of the long-running debate about the impact on productivity of information and communications technologies (ICT).

This began with the famous "Solow paradox," the observation in 1987 by Robert Solow that "you can see the computer age everywhere but in the productivity statistics" [3]. Various explanations for this paradox were identified. Whatever the explanation, later studies in growth accounting (breaking down productivity changes into components attributable to labour inputs, capital inputs, and technical progress) demonstrated a much greater effect. Thus a study by Jorgenson et al. [4] concluded that ICT was responsible for 50% of US labour productivity growth between 1995 and 2000 and 33% between 2000 and 2005. Timmer and Van Ark [5] concluded that higher ICT investment explains more than half the US advantage over Europe in labour productivity growth from 1995 to 2001. A recent review of the empirical literature concludes that the productivity effect of ICT is significant, positive and increasing. A US/Europe differential is found in aggregate, but this may be due to lower levels of ICT investment in Europe [6, p. 117].

Breaking down the impact of ICT into the contributions of its components such as mobile communications presents enormous problems, especially if the different ICT inputs, notably information processing and communications, are complementary, as has been suggested. There is, however, a more direct way of establishing the effect on the economy of communications, and mobile communications in particular, to which we now turn.

13.4.1 Estimating the macroeconomic effects of the diffusion of communications services, including mobile communications

One of the debates about ICT focusses upon whether it is a "general-purpose technology" or an "enabling technology," with the characteristic that it affects a multitude of economic activities, all of which it makes more productive.[4] This opens up the possibility that it can have considerable "spillover effects" throughout the whole economy, which are not captured by the analyses described above. Since our focus is a narrower one on spectrum use, it is not necessary to decide this broader question relating to ICT as a whole.

In relation to broadband it is a common and everyday observation that few acts of consumption and production by those with access to broadband are wholly immune from its effects on their lives. In 2013, three-quarters of those connected to broadband were using a mobile connection, and this proportion will rise further as broadband connections grow in the less developed world, which has much more limited access to fixed networks [9].

Since then a process of enumeration of wider effects of broadband has taken place. A typical classification includes the following:

- Enhanced speed and quality of information flows: sometimes it is suggested that the combination of more information processing and faster communications necessary to deliver the benefits, with one alone producing less spectacular results.
- Better access to markets: due to lower barriers to entry, an increase in the geographical scope of markets (the "death of distance"), better job matching, better access to customers via the Web etc.[5]

---

[4] The standard precursor of ICT as a general-purpose technology is electricity. See [7, 8].
[5] For an illustration of how this takes effect with voice communications, see the study by Jensen [10] describing how fishing vessels in south India, before landing their catch, can call the shore to establish in which port the best prices can be found.

- New business processes and organizational structures: better stock control, quicker contracting, just-in-time production etc. (For example, a large grocery company operates in the US and in several Latin American countries. It is reported that the lack of reliable broadband in one of the Latin American countries leads to an entirely different approach to logistics than is applied in the USA.)
- More innovation in general: made possible by the availability of new communications services; examples can be multiplied – social networks being a particularly significant one.

This is a plausible story, and some effort has gone into testing it. One strategy would be to try to identify the impacts of mobile voice and broadband in each sector, and aggregate them. This would be very complex and the results inevitably contestable. An alternative is to adopt a cross-section approach: to collect data on the diffusion of voice and broadband services in a sample of countries, and test the hypothesis that higher diffusion rates are associated with higher levels of GDP.

The method is more complicated when the sample includes countries some of which rely virtually exclusively on wireless networks and others where both fixed and wireless voice and data services are available. In the latter case, for our purpose, some partitioning of effects between spectrum-using and non-spectrum-using services is required.

The simple modus operandi of such studies is illustrated by an influential 2009 World Bank study of developing countries [11]. It involved estimating the following regression equation:

The average growth rate of per capita GDP between 1980 and 2006 was used as the dependent variable and regressed onto the following variables, selected as representative of conditioning variables in the growth literature:

- Per capita GDP in 1980
- Average ratio of investment to GDP between 1980 and 2006
- Primary school enrolment rate in 1980 (a proxy for human capital stock)
- Average penetration of broadband and other telecommunications services between 1980 and 2006 for developed and developing countries (a proxy for technological progress and the focus of the analysis)
- Dummy variables for countries in the Sub-Saharan Africa, Latin America and Caribbean regions.

The problem with this procedure is that it does not unpack the rather complex two-way supply-and-demand interactions which take place between a sector and the macro-economy. Other studies use more sophisticated structural models, but, as shown below, the range of results is very wide.

Waverman et al. [12] led the way with an analysis of the mobile sector, which suggested that differences in mobile voice diffusion accounted for a significant part of GDP differences across a wide variety of countries. A later study of this technology by Gruber and Koutroumpis [13] found a significantly smaller effect.

Table 13-3. Estimates of the effect on growth of increased penetration

Authors	Countries	Effect on growth of 10% additional broadband penetration
Czernich et al. [14]	OECD, 1996–2007	0.9–1.5%
Katz and Avila [15]	24 Latin American and Caribbean countries	0.2%
Gruber and Koutroumpis [13]	EU15, 2003–2006	0.26–0.38%
OECD [16]	EU countries, 1980–2009	1.1%

Table 13-4. Estimates of the effect on growth of increased penetration of telecommunications services

% increase in economic growth per 10% increase in penetration, in:	Fixed	Mobile	Internet	Broadband
High-income countries	0.4	0.6	0.8	1.2
Low-income countries	0.7	0.8	1.1	1.4

Source: [11].

Similar studies have been made of the impact on growth of a 10% increase in the diffusion of broadband. A summary of some recent results is shown in Table 13-3. The results of the above-noted World Bank study are shown in Table 13.4.

If these results were confirmed as applying exactly to a country which started from a zero base of broadband penetration, and achieved an annual increase in broadband penetration of 10% per year for ten years, then that country's GDP would grow 1.2% faster than it would otherwise in each of those ten years. However, a number of problems of interpretation arise in connection with these results:

- The samples of countries in some studies are fairly heterogeneous, with respect to income per head and endowment of fixed networks. In some countries, effectively all broadband is wireless. In others, mobile broadband has been superimposed upon widely diffused fixed broadband services. This is likely to influence patterns of use, as between, say, consumption and productive uses.
- As noted, most studies to date have focussed on fixed broadband, whereas our interest is in broadband delivered by wireless. It is not clear that the two modes of delivery will have similar economic externalities, especially in countries in which the population has access to both, where mobile broadband may be principally associated with use for

consumption purposes – which may, however, generate noneconomic externalities (see section 11.6 above).
- Most of the work is done by agencies associated with the telecommunications sector, which may have a vested interest in strong results. This may lead to a selection bias, in consequence of which only positive results are disseminated.
- There may also be issues about the thoroughness of pre-publication review. Only in a minority of cases (notably Czernich et al. [14]) have the results appeared in academic journals in which refereeing is the rule.[6]
- There are almost certainly missing variables in the estimated equations, ranging from investment in other activities (see below) to – closer to home – the presence or absence of adequate investment in digital literacy.
- It is worth noting that the communications sector is not alone in being identified as an engine of growth. Equivalent claims are made in respect of transport [17], infrastructure in general [18], and education and health [19]. More generally, the literature on endogenous growth theory (see, for example, Barro [19], Aghion and Howitt [20]) identifies a role for investment in public capital to augment long-term growth rates. Angelopoulos, Economides, and Kammas [21] find a positive effect of public investment on growth rates in OECD countries over an extended period. If broadband diffusion goes hand in hand with other developments, it may be hard to separate the effects. Moreover, if estimates of all the effects were applied at once, changes in GDP levels might be "explained" several times over.

13.5 Conclusion

The direct estimates of the contribution of mobile communications to welfare and gross domestic product cited in Tables 13-1 and 13-2 above are quite small: equivalent to 1 to 3% of GDP. At the same time we find some high estimates of total impact (including externalities) on GDP of the diffusion of mobile voice and of fixed broadband. Thus, if the estimates of Czernich et al., which suggest that a 50% diffusion of fixed broadband (a level now far surpassed in many countries) would lead to an increase in GDP of 4.5 to 7.5%, apply also to mobile broadband, its spread could have a major impact on GDP. More research is needed to investigate this effect.

---

[6] This does not exclude the possibility of forms of refereeing in other outlets.

13.6 Annex

Table 13-5. Various estimates of the economic consequences of spectrum use

Year	Region	Form of economic value	Industry or action	% of GDP
2012[7]	Australia	Direct and indirect economic contribution	Mobile communications	0.48%
2014[8]		Household consumption	Mobile broadband	0.93%
2011[9]	UK	Economic welfare (consumer and producer surplus)	Mobile communications	1.99%
			Wi-Fi	0.12%
			TV broadcasting	0.51%
			Radio broadcasting	0.2%
			Microwave links	0.22%
			Satellite links	0.24%
			Private mobile radio	0.08%
2008[10]	US	Productivity increase	Mobile communications	1.47%
2007[11]	Japan	Direct and indirect economic contributions	Mobile industries	1.68%
2010[12]	OECD area	Productivity increase	Mobile telecommunications	0.39%
2009[13]	Australia	Economic welfare (consumer and producer surplus)	Mobile broadband by UHF	0.089%
2011[14]	Latin America	Consumer surplus	Mobile broadband at 700 MHz	0.068%
2013[15]	European Union	Consumer surplus	Mobile services	2.04%
			Terrestrial broadcasting	0.37%
			Satellite communications	0.13%
			Private mobile radio	0.14%
			Civil aviation services	1.30%
2014[16]	USA	Consumer surplus	LSA[17] from 100 MHz to 3.5 GHz	0.08%
		Consumer surplus	LSA from 50 MHz to 2.3 GHz	0.04%

Source: Adapted, with additions, from New Approaches to Spectrum Management, OECD Digital Economy Papers No. 235, 2014, pp. 47–49. For further details, see the original, available at www.oecd-ilibrary.org/science-and-technology/new-approaches-to-spectrum-management_5jz44fnq066c-en

---

[7] See [22]. [8] See [23]. [9] See [24]. [10] See [25]. [11] See [26]. [12] See [27].
[13] See [28]. [14] See [29]. [15] See [30]. [16] See [31]. [17] Licensed shared access.

References

[1] Department for Business, Innovation and Skills (BIS), Department for Culture, Media and Sport (DCMS), and Analysys Mason (2012), "Impact of Radio Spectrum on the UK Economy and Factors Influencing Future Spectrum Demand."

[2] Australian Mobile Telecommunications Association (AMTA) and Deloitte Access Economics, "Mobile Nation: The Economic and Social Impacts of Mobile Technology" (2013), at www.amta.org.au/pages/State.of.the.Industry.Reports.

[3] R. Solow, "We'd Better Watch Out," *New York Times Book Review*, July 12, 1987, 36.

[4] D. Jorgenson et al., *A Retrospective Look at the US Productivity Growth Resurgence*, Federal Reserve Bank of New York, 2007.

[5] M. Timmer and B. van Ark, "Does Information and Communication Technology Drive US/EU Productivity Growth Differentials?" (2005) 57(4) *Oxford Economic Papers* 693.

[6] M. Cardona, T. Kretschmer, and T. Strobel, "ICT and Productivity: Conclusions from the Empirical Literature" (2013) 25 *Information Economics and Policy* 109.

[7] T. Bresnahan, "General Purpose Technologies," in B. Hall and N. Rosenberg, eds., *Handbook of the Economics of Innovation*, Vol. 2, Amsterdam: Elsevier North Holland, 2010, 761.

[8] J. Jovanovich and P. Rousseau, "General Purpose Technologies," in P. Aghion and S. Durlauf, eds., *Handbook of Economic Growth*, Vol. 1B, Amsterdam: Elsevier North Holland, 2005, 1181.

[9] Broadband Commission, "The State of Broadband 2014: Broadband for All," ITU/UNESCO, 2014, 96.

[10] R. Jensen, "The Digital Divide: Information (Technology), Market Performance and Welfare in the South Indian Fisheries Sector" (2007) 122(3) *Quarterly Journal of Economics* 879.

[11] C. Qiang et al., "The Economic Impact of Broadband," in World Bank, *Information and Communication for Development*, Washington, DC, 2009, 35, available at http://siteresources.worldbank.org/EXTIC4D/Resources/IC4D_Broadband_35_50.pdf.

[12] L. Waverman et al., "The Impact of Telecoms on Economic Growth in Developing Countries," in *The Vodafone Policy Paper Series, Number 2*, London, 2005, 10.

[13] H. Gruber and P. Koutroumpis (2010), "Mobile Telecommunications and the Impact on Economic Development," draft for Fifty-Second Economic CEPR Policy Panel.

[14] R. Czernich et al., "Broadband Infrastructure and Economic Growth" (2011) 121 (552) *Economic Journal* 505.

[15] R. Katz and J. Avila, "Estimating Broadband Demand and Its Economic Impact in Latin America" (2010), *Proceedings of the 4th ACORN-REDECOM Conference*, Brasilia, May 14–15.

[16] OECD, *Economic Impact of Internet/Broadband Technologies*, DSTI/ICCP/IE (2011) i/Rev1, Paris, 2011.

[17] D. Banister and Y. Berechman, "Transport Investment and the Promotion of Economic Growth" (2001) 9(3) *Journal of Transport Geography* 209.

[18] A. Munnell, "Infrastructure Investment and Economic Growth" (1992) 6(4) *Journal of Economic Perspectives* 189.
[19] R. Barro, "Government Spending in a Simple Model of Endogenous Growth" (1990) 95(8) *Journal of Political Economy* S103.
[20] P. Aghion and P. Howitt, *Economics of Growth*, Cambridge, MA: MIT Press, 2009.
[21] K. Angelopoulos, G. Economides, and P. Kammas, "Tax-Spending Policies and Economic Growth: Theoretical Predictions and Evidence for the OECD" (2007) 23 *European Journal of Political Economy* 885.
[22] Australian Mobile Telecommunications Association (AMTA) and Deloitte Access Economics (2013), "Mobile Nation: The Economic and Social Impacts of Mobile Technology," www.amta.org.au/pages/State.of.the.Industry.Reports.
[23] Centre for International Economics, "The Economic Impacts of Mobile Broadband on the Australian Economy, 2006–2013" (2014).
[24] Department for Business, Innovation and Skills (BIS), Department for Culture, Media and Sport (DCMS) and Analysys Mason (2012), "Impact of Radio Spectrum on the UK Economy and Factors Influencing Future Spectrum Demand," at www.analysys mason.com/About-Us/News/Insight/Economic-value-ofspectrum-Jan2013/#. UbhPiNLVBYR.
[25] CTIA-The Wireless Association and R. Entner (2008), "The Increasingly Important Impact of Wireless Broadband Technology and Services on the U.S. Economy," http://files.ctia.org/pdf/Final_OvumEconomicImpact_Repor t_5_21_08.pdf.
[26] InfoCom Research (2007), "An Analysis on Ripple Effect of Mobile Phone Services on the Japanese Economy," www.icr.co.jp/press/press20070824_bun seki.pdf (in Japanese).
[27] H. Gruber and P. Koutroumpis (2010), "Mobile Telecommunications and the Impact on Economic Development," draft for Fifty-Second Economic CEPR Policy Panel, at www.cepr.org/meets/wkcn/9/979/papers/Gruber_Koutroumpis.pdf.
[28] Australian Mobile Telecommunication Association (AMTA), "Spectrum Value Partners and Venture Consulting (2009), "Optimal Split for the Digital Dividend Spectrum in Australia," at www.amta.org.au/files/SVP.Report.Executive.Sum mary.pdf.
[29] GSM Association (GSMA), Asociación Iberoamericana de Centros de Investigación y Empresas de Telecomunicaciones (AHCIET) and Telecom Advisory Services (2011), "Economic Benefits of the Digital Dividend for Latin America," at www.gsma.com/latinamerica/economic-benefits-of-the-digi tal-dividend-for-latin-america.
[30] GSM Association (GSMA) and Plum Consulting (2013), "Valuing the use of Spectrum in the EU," at www.gsma.com/spectrum/valuing-the-use-of-spec trum-in-the-eu.
[31] GSM Association (GSMA), Deloitte and Real Wireless (2014), "The Impact of Licensed Shared Use of Spectrum," at www.gsma.com/spectrum/the-im pact-of-licensed-shared-use-of-spectrum.

14 Where next?

The challenge for the regulator remains the same as always – to maximize the value of the use of radio spectrum through ensuring that the highest-value users have access to the spectrum and that the interference between them remains controlled to an optimal level. However, the solution to this challenge is constantly changing as usage patterns change and new tools to manage users develop, often based on changing technology. In this final chapter we look at the key trends relating to spectrum usage and suggest where we think spectrum management will evolve in the coming decades.

14.1 Trends

14.1.1 Trends in usage

There are hundreds of different uses of the radio spectrum and trends and changes in many of these. However, where the users require relatively small amounts of spectrum, or spectrum for which there is little competition (e.g. satellite use of frequencies above 20 GHz) then these have limited impact on spectrum management approaches. Here we discuss what we believe are some of the most important trends relating to those uses that generate the highest economic value (broadly cellular and broadcasting), those that compete for the most congested spectrum and those where innovation might be most likely to occur. The trends are not presented in any particular order.

Trend 1: growing importance of unlicensed spectrum We are using Wi-Fi and Bluetooth ever more. Home Wi-Fi networks now routinely have tens of devices connected to them whereas a decade ago they might have had one or two. Monitoring devices are slowly appearing around the home, such as Wi-Fi-connected security devices, and many more are being shown at trade shows. Other systems, such as the home energy system, may make use of a separate network within the home to connect gas and electricity meters, in-home displays, and eventually smart appliances. Out of the home we are increasingly looking to log onto Wi-Fi hotspots wherever we find ourselves. Bluetooth accessories are proliferating, with many health- and fitness-oriented devices being developed for wearing and devices like Google Glass being tested. As this usage grows so will the possibilities of interference, while at the same time our tolerance for not being connected will fall. The use of 5 GHz spectrum will offer some relief but regulators may need to become more inclined towards making bands unlicensed than they are at present.

Trend 2: ubiquitous connectivity will become ever more important Once connectivity is widespread then behaviors change. People make greater use of cloud servers and plan ahead less on the assumption that they can retrieve the information they need as and when it is needed. Less material is stored locally, enabling devices without hard disks. All of this makes ubiquitous connectivity ever more important, fueling a circle of change. This is perhaps less about rural coverage and more about ensuring that all of the devices that a person is carrying can connect wherever they happen to be.

Trend 3: ever higher data rates will become less important Each new generation of mobile system (e.g. 2G, 3G, 4G) has been predicated on approximately a tenfold increase in data rates compared to the previous generation. This has led to requirements for ever broader bandwidths and ever more mobile spectrum. But we are now reaching a point where many homes and users have more than adequate data rates for all that they could wish to achieve. It is hard to envisage why we might want yet another tenfold increase. This is causing a rethink as to whether fibre to the home is a necessary solution, and it will bring into question whether data rates moving towards 1 Gbit per second are necessary for 5G. This in turn will change the demand for spectrum for future mobile data systems. Coupled with the trend toward ever more use of Wi-Fi it may be that the seemingly insatiable desire for spectrum for mobile usage will finally come to an end.

Trend 4: the Internet of Things will grow rapidly The Internet of Things (IoT) is predicted to grow from perhaps 1 billon connected devices in 2015 to over 50 billion by 2025. This rapid growth will require some new spectrum, which regulators have not yet identified, and will provide a new source of value from the radio spectrum.

Trend 5: the categorization of usage into fixed, mobile, and broadcasting will become of less relevance At present regulators divide usage into fixed, mobile or broadcasting. But these are increasingly converging. Broadcast and mobile systems will tend to share platforms and spectrum in the future and broadcast-only solutions will increasingly be switched off.

14.1.2 Trends in spectrum management tools

A range of new tools will become available for spectrum management. These will include the following trends.

Trend 6: dynamic spectrum sharing will become commonplace As white space deployments grow and sharing moves into other bands, the concepts of shared opportunistic access will become more widely accepted and it will be considered normal for virtually all spectrum bands to be shared in some manner.

Trend 7: regulators will stop predicting interference and start monitoring it The complexity of predicting what interference might occur will continue to grow while new tools will become available to rapidly fix interference where it does occur. As a result, regulators will increasingly deploy systems up to a point at which interference is observed and then set this as the license conditions. New ways of monitoring interference will be needed to allow for real-time reporting across a range of users and uses.

Trend 8: regulators will seek more ways to encourage innovation Regulators often aspire to stimulate innovation but rarely achieve it. Innovation is becoming ever more important in government agendas and regulators will start looking at mechanisms to provide favored allocations for new approaches and other novel ways to encourage innovation.

Trend 9: receiver performance will become an ever-growing issue and regulators will eventually move to control receiver performance as well as transmitter performance The performance of receivers often prevents or limits new technologies or uses in neighboring bands. Regulators will increasingly realize that some specification of expected receiver performance is needed in order to effectively manage spectrum.

14.2 Our agenda to improve spectrum use

Throughout this book we have discussed areas where current approaches to management and use of spectrum fall short. In such cases we have put forward suggestions as to how they might be improved. The five most important proposals of this kind are set out below.

1. Move to a position where (almost) all licenses are shared The coexistence of increasing demand for spectrum and evidence that many valuable frequencies are underused has been one of the drivers, over the last five years, for more spectrum sharing. This process has been assisted by the development of new real-time technologies for dynamic spectrum sharing which allow multiple users to coexist. These methods have augmented earlier temporal and geographic sharing of bands and the limited use of unlicensed spectrum.

As noted in section 9.9, it is time for these possibilities to be reflected more fully in rights of access to spectrum by the replacement of exclusive licenses by arrangements which allow access to multiple users, possibly on a hierarchical basis which gives some users priority over others. The result to be expected is much greater flexibility in use of spectrum and lower prices of access to it. This could be accomplished by a process of progressively replacing exclusive licenses with less restrictive alternatives, introduced in ways which manage the associated risks. One effect might be the emergence of intermediaries which offer flexible access to multiple clients. We recommend in the future a brisk increase in the number of licenses recast in this way, even if in practice some of these will continue to be exclusive.

There is mounting evidence that inefficient use of public spectrum is increasing relative to commercial spectrum. Chapter 12 outlines a strategy for dealing with this, but its success hangs in the balance in many countries. Public/commercial sharing of spectrum (see sections 9.7–9.8) may in some countries be a more palatable means of switching spectrum to more efficient uses.

2. Link licenses to interference generated rather than power transmitted Current licenses are not suitable for a world where usage can be changed and bands repurposed. There have been many examples of this, from the interference caused by Nextel in US public safety systems in the 1990s through to the more recent cases of LightSquared in the US and TV interference concerns in white space. The basic problem is that a license that specifies the amount of power a transmitter is allowed to emit does not fully constrain the interference caused since this is also determined by the density of transmitters and other factors such as their height. A better approach is to specify in the license the interference that can be caused and then allow the license holders to configure their transmitters in such a way that this interference level is not exceeded. Studies such as Ofcom's Spectrum Usage Rights (see section 10.2) have pioneered more suitable license formats that achieve these objectives while being practical to implement. However, existing players have generally resisted their introduction due to increased complexity and the problem that the benefits most immediately fall on the newer entrants rather than established companies. Regulators should look at the wider benefits for usage of radio spectrum rather than the needs of incumbents and adopt licenses specified in interference terms as the default across all spectrum bands.

3. Manage receiver performance There are two aspects to the performance of a radio receiver – its ability to decode the wanted signal and its ability to reject unwanted or interfering signals. While the former is typically a core part of radio design, the latter is more problematic. Manufacturers may not know what interfering signals will be experienced either now or in the future and have strong incentives to design the lowest-cost receiver by taking a low expectation of interference. If such poorly performing receivers are widely deployed, for example in millions of TV sets, this can cause the regulator to prevent the introduction of nearby spectrum usage that might impact these receivers, despite the fact that they are substandard. Knowing that this outcome is likely allows receiver manufacturers to design to ever-lower standards, resulting in a "too-big-to-fail" problem. To some degree this problem is linked to license conditions (see above), in that a clearer specification of the interference that spectrum neighbors could cause would clarify the receiver requirements. There are many issues here associated with national licensing versus global manufacturing, the role of standards bodies, and more that we discuss in detail in section 10.3. It is clear that this is a serious problem that is progressively causing greater delays in optimal use of spectrum and needs addressing. Regulators need to work together on a global basis, perhaps through the ITU or similar bodies, to ensure that receivers have appropriate specifications and that these are implemented.

4. Move away from previous ITU labels to a new categorization Bands are currently divided into categories such as "fixed," "mobile," and "broadcast" at international level and have been for nearly a century. However, the boundaries between these uses have been increasingly blurring. For example, many now view broadcast content on mobile devices, downloaded via cellular or Wi-Fi. There are mechanisms such as "broadcast modes" within cellular networks (see section 11.3) which allow broadcasting to be performed from those cellular networks. The labels were originally intended to facilitate interference management by placing the highest-power networks (broadcasting) into some bands and lower-power networks (mobile) into others. But as discussed above this method of managing spectrum is suboptimal. From a management perspective it is the interference levels generated, rather than the service deployed, that is most relevant. The categorization also tends to limit innovation by persuading users that they cannot, for example, change use to a mobile application in a band that has been categorized as broadcast, whereas in practice this particular example would likely work acceptably. A better division would be "low," "medium," and "high" interference levels, ideally linked to specific limits and new license types. The idea of reclassification should be discussed internationally by spectrum managers and others, options considered, and a new approach adopted at ITU level and cascaded down to regional and national plans.

5. Reconsider regional and global spectrum management There is greater scope for regional collaboration if appropriate frameworks can be found that do not limit innovation. As discussed in section 1.6, regional bodies should study their role and look for where they can add additional value. This is an issue of fundamental importance which deserves careful and disinterested study at a very high level.

14.3 In conclusion

Spectrum management has come on a long journey through its 110-year history. Technological developments have led to rapidly growing demand and also to innovative options for using spectrum based around sharing and reactive assignment.

These developments have exposed the weaknesses of the legacy command-and-control approach to spectrum management. Initially this led to the development of market-based management in the past 25 years, initially focussed on auctions but later extended to secondary markets in spectrum and also to spectrum pricing.

With the notable exception of auctions, the market-based methods have not been widely applied, and have not yet delivered all of the expected gains. The world where regulators step back and allow the market to function still looks to be in the distant future. Making the best use of spectrum appears to need a mix of new technology and new regulation. The most prominent area of application of this combination presently resides in spectrum sharing, based on techniques such as dynamic spectrum access. Other developments which have strong potential are the implementation in licenses of new ways to control interference, and the extension of regulation into domains such as receiver performance.

As spectrum becomes an ever more critical component of our lives the difficulty of making such changes grows. The risk remains that cautious regulators forever remain one step behind technology and need. This book has set out how we believe they and others can best act to put the airwaves to work for the benefit of the economy and society. The prize is enormous.

About the authors

Martin Cave

Martin Cave is a regulatory economist who has worked extensively on telecommunications and spectrum issues. He is a visiting professor at Imperial College Business School and an Inquiry Chair at the UK Competition and Markets Authority. In 2010–2011 he held a BP Centennial Chair at the London School of Economics, from 2001 to 2009 he was a professor at Warwick Business School and before that professor of economics at Brunel University. He was awarded bachelor's, master's and doctoral degrees in economics from Oxford University.

He has written extensively on regulation of the communications sector, including academic papers on the role of *ex ante* regulation, investment incentives, the structure of the mobile sector, broadcasting issues, spectrum markets, spectrum management, and spectrum pricing. He is coeditor of the *Handbook of Telecommunications Economics* (Vols. 1 and 2, Elsevier, 2002 and 2006), and coauthor of *Understanding Regulation* (Oxford University Press, 2011).

He carried out a Review of Radio Spectrum Management in 2002 and an Independent Audit of Major Spectrum Holdings in 2006 for the UK government. He has participated in similar strategic spectrum reviews for the governments of Australia and Canada. He has undertaken studies for the European Commission, IADB, ITU, and OECD, and advised spectrum regulators and firms in many countries in Africa, Asia, Australia, Europe, and North and South America on the use of markets, assignment policy in particular bands, spectrum pricing and valuation, and other matters. He was a member of the Spectrum Advisory Board of the UK regulator Ofcom from 2004 to 2007 and is a member of an advisory group to the Irish telecommunications and spectrum regulator.

William Webb

William Webb is the CEO of the Weightless SIG, the body developing the Weightless Standard, and an independent consultant on wireless technology and spectrum regulation. He was one of the core design team involved in Weightless and instrumental in establishing the SIG.

Prior to this William was one of the founding directors of Neul, a company developing machine-to-machine technologies and networks, and before that a director at Ofcom, where he managed a team providing technical advice and performing research across all areas of Ofcom's regulatory remit. He also led some of the major reviews conducted by

Ofcom, including the Spectrum Framework Review, the development of Spectrum Usage Rights, and most recently cognitive or white space policy. Previously, William worked for a range of communications consultancies in the UK in the fields of hardware design, computer simulation, propagation modeling, spectrum management, and strategy development. William also spent three years providing strategic management across Motorola's entire communications portfolio, based in Chicago.

William has published 13 books including *Understanding Weightless*; over 100 papers; and 18 patents. He is a Visiting Professor at Surrey University, a member of Ofcom's Spectrum Advisory Board (OSAB), and a Fellow of the Royal Academy of Engineering, the IEEE and the IET, where he was President for the 2014–2015 session. William has a first-class honors degree in electronics, a PhD, and an MBA. In his spare time he is a keen cyclist, having ridden from Land's End to John O'Groats and the route of the Tour de France, and completed the grueling Cent Cols Challenge four times, riding over 100 Alpine and Dolomite mountain passes in 10 days.

Index

2G, 7
3D-TV, 210
3G, 7, 78, 82, 99
3GPP, 35
4G, 7, 16, 143
4G cellular, 39
5G, 207
ACL, 35
ACMA, 57, 123, 139
activity rules, 77
administrative pricing, 55, 128
advanced wireless services. *See* AWS
aeronautical, 4, 137
AGC, 37, 177
AIP, 137
allocation, 44
    efficient, 44
amateur, 8
antenna, 24, 175, 176, 226
appliances, 252
at800, 39
ATC, 198
auction, 9, 51, 68
    3G, 78
    ascending clock, 82
    first-price, 70
    first-price sealed-bid, 74
    format, 72
    hybrid, 84
    open, 70
    revenue share, 83
    SAA, 77
    sealed-bid, 70
    second-price, 70
    second-price sealed-bid, 75
    sequential, 84
auction guide, 101
audits, 236
Australia, 97, 123, 193, 214
AWS, 198, 218

backhaul, 5, 166
band managers, 115
bandwidth, 24
bank guarantees, 96, 100
BAS, 198
BBC, 139
beacons, 164
beauty contest, 51, 128
benefit, 45
benefits
    external, 219
BFWA, 81
blocking, 37
Bluetooth, 5, 149, 252
Broadband Commission, 215
broadband fixed wireless access. *See* BFWA
broadcasting, 7
    terrestrial, 207
building penetration, 31
bursty, 177
business model, 57
business radio, 8

cable, 39, 176
    losses, 176
Canada, 119, 130
capacity, 226
cartel, 99
C-band, 198
CENELEC, 203
CEPT, 121
cloud servers, 253
Coase, 10, 49
codecs, 211
codes of conduct, 171
COGEU, 124
cognitive pilot channel, 165
command and control, 10, 48
Commercial Spectrum Enhancement Act, 235
common value, 63
commons, 149
competition, 103, 117
competition law, 105
compression, 42, 212
ComReg, 142

congestion, 153
conjoint analysis, 221
connectivity, 253
consumer surplus, 242
co-ordinated conduct. *See* cartel
coverage, 216
coverage contours, 169
CW, 177

DAB, 134
database, 156, 188
DECT, 157
default, 96
defense, 231
defragmentation, 187
Demsetz, 49
Denmark, 121
Department of Defense, 235
deployment density, 193
deposit, 95, 100
deprival value, 132
Digital Agenda for Europe, 16
Digital Dividend Review, 220
digital switch-over, 211
directional, 25
directives, 203
dispute resolution, 115
DoD, 184, 237
drones, 4
DTT, 8, 132, 227
ducting, 33
duty cycle, 150, 155, 171
DVB, 200
DVB-T, 164
DVB-T2, 133, 211

earth observation satellite, 6
ECC, 155
economic benefits, 11
economic welfare, 241
EIRP, 25
electromagnetic wave, 24
eMBMS, 214
EMC, 203
ETSI, 203
Europe, 73
exclusive use, 188
extensive margin, 42
externalities, 115

fast fading, 30
FCC, 73, 122, 160, 176, 218
FDD, 35
filters, 36, 198
first-price sealed-bid, 67
fixed links, 5

foreclosure, 104
forfeiture, 103
frequency hopping, 169
FSS, 184

GAA, 184
Galileo, 6
Gaussian, 27
GDP, 208, 241, 243
Geneva, 15
geo-location, 166, 188
Germany, 216
Google Glass, 252
GPS, 6, 38, 166, 199
GSM, 24
GSMA, 199
guard band, 35
Guatemala, 54, 119

harm claim thresholds, 197
harmful interference, 13
harmonics, 27
harmonization, 120
Hata, 29, 175, 178
hertz, 24
hidden-terminal problem, 163
hoarding, 241
Hong Kong, 83, 84, 99
HTHP, 209, 214

ICT, 215
IEEE, 259
IET, 259
incentive auction, 226
incentive pricing, 186
indoor TV, 183
information memorandum, 101
Inmarsat, 6
innovation, 5, 22, 113
intensive margin, 42
interference, 24, 33, 114, 169, 193, 254
intermodulation, 27
invitation stage, 94
IoT, 253
iPhone, 217
IPTV, 227
Ireland, 142
ISD, 214
ISM, 150
isotropic, 25
ITU, 9, 15, 48, 189, 215

Jorgenson, 245

laissez-faire, 40
L-Band, 197

liberalization, 52
licence-exempt, 5, 149
lifeboats, 137
light licensing, 155
LightSquared, 38, 198, 255
listen-before-talk, 152
Lithuania, 121
logistics, 94
log-Normal distribution, 175
LSA, 124, 167
LTE, 7, 39, 184
LTLP, 214

M2M, 5, 171, 227
map colouring, 209
maritime, 8, 137
markets, 42, 49, 114
MCL, 32, 179, 182, 183
mechanism design, 63
MED, 123
merger, 99
meteorological, 6
Mexico, 119
MFN, 213
microwave ovens, 150
Milgrom–Wilson activity rule, 77
military, 164
MIMO, 39
mixer, 37
MNO, 124
mobile TV, 210
monopoly, 57
MPEG, 212
multiplex, 177, 211

NAB, 237
National Broadband Plan, 186
National Science Foundation, 124
national security, 233
National Telecommunications and Information Administration. *See* NTIA
NATO, 16
new public management, 238
New Zealand, 73, 76, 97, 123
Nextel, 198, 255
Nigeria, 83
Noam, 189
noise, 226
noise floor, 163
NRA, 11, 216
NTIA, 73, 186

OECD, 215
Ofcom, 12, 39, 57, 132, 137, 143, 164, 176, 194, 221, 258
omnidirectional, 25
opportunity cost, 56, 129, 132

option value, 56
OSAB, 259
Ostrom, 152

Part 15, 159
PCAST, 185, 200
PFD, 194
pixel, 170, 175
Plum Consulting, 133
PMR, 8, 122
PMSE, 3
polarization, 178
politeness protocols, 150, 171
power, 24
power control, 171
PPDR, 5, 226
pre-qualification stage, 95
primary, 162
private commons, 47, 109
private value, 63
producer surplus, 242
productivity, 244
propagation, 29
  free-space, 28
  real-life, 29
propagation models, 169, 178
property rights, 114
protection ratios, 179
PSB, 215
public sector, 231
public-service broadcasting, 208
PVRs, 213

R&TTE, 203
radar, 4, 162, 198
radio altimeters, 4
radio astronomy, 162
receiver performance, 197, 254
receiver protection, 200
receiver standards, 201
receivers, 38, 255
recess, 96
refarming, 57, 234
regulator, 40
reserve price, 82, 100, 128
revenue equivalence, 70
RFID, 158
RSPG, 121, 203
RSPP, 16, 203

SAS, 184
satellite, 6, 39, 162
satellite earth station, 198
scarcity, 128
SDARS, 198
secondary, 162
sensing, 163, 187

Index

service neutrality, 52
SFN, 133, 213
shadow prices, 236
Shannon, 26
shared access
   unlicensed, 168
sharing, 47, 237, 253
   vertical, 47
SIM, 9
simulcasting, 210, 213
SiriusXM, 198
SLC, 119
slow-fading, 30
snake-in-the-grass, 77
social goals, 11
social value, 220
Solow paradox, 244
South Korea, 210
spectrum
   usage, 162
spectrum brokers, 115
spectrum cap, 52, 97, 105, 119
spectrum commons, 5
spectrum crunch, 10
Spectrum Framework Review, 259
spectrum inventory, 237
spectrum manager, 63
spectrum mask, 26
spectrum registries, 114
Spectrum Relocation Fund, 235
spectrum rights, 114
spectrum sharing, 254
spectrum trading, 53, 113
spectrum usage rights, 194, 259
speculation, 95
spurious emissions, 27
standards, 200
stated preference, 221
STUs, 123
subsidy, 220
substitution, 45
SURs, 53, 202
Sweden, 216
Switzerland, 84, 121

tax revenue, 73
TDD, 36, 194
technological neutrality, 52
testbed, 186
Thailand, 130
time division duplex. *See* TDD
TNR, 122
tragedy of the commons, 152
tropospheric ducting, 175
TTP, 202
TUFs, 123
TV white space, 174, 198, 219
TVWS, 124
type 1 errors, 117
type 2 errors, 117

UHDTV, 210
UHF, 207
UN, 9, 17
unassigned, 9
UNESCO, 215
unilateral conduct, 99
unlicensed, 5, 149, 252
unmodulated, 24
US, 73, 227
use-it-or-lose-it, 119
utilization, 162
UWB, 159

Vickrey mechanism, 110

waivers, 96
WCS, 198, 218
Weightless, 227, 259
Weightless SIG, 258
white space, 253, 255
Wi-Fi, 5, 149, 252
willingness to pay, 221
windfall gains, 120
winner's curse, 64
wireless cameras, 6
wireless microphones, 6
WRC, 15, 227